(No Model.)

3 Sh

N. TESLA.
SYSTEM OF ELECTRICAL POWER TRANSMISSIO

No. 511,560.

Patented Dec. 26, 1893.

JAN — 2016

D1252626

Fig. 1

Generator

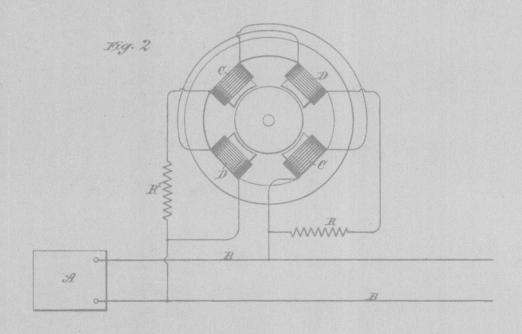

Fig. 2

INVENTOR

Nikola Tesla

BY

Duncan, Curtis & Page

ATTORNEYS.

THE
TRUTH ABOUT
TESLA

For my two favorite patent agents,
Rob Pilaud and Filip Vanevski, whose work
I would never describe as imprecise and ignoble.

THE
TRUTH ABOUT
TESLA

THE MYTH OF THE LONE GENIUS
IN THE HISTORY OF INNOVATION

CHRISTOPHER COOPER
FOREWORD BY MARC J. SEIFER, PH.D.

Race Point
PUBLISHING

Quarto is the authority on a wide range of topics.

Quarto educates, entertains and enriches the lives of
our readers—enthusiasts and lovers of hands-on living.

www.quartoknows.com

Illustrations (pages 68, 72, 76, 79, 81, 83, 89, 90, 91, 95, 117, 123, 141)
© 2015 Quarto Publishing Group USA Inc.

First published in the United States of America in 2015 by
Race Point Publishing, a member of
Quarto Publishing Group USA Inc.
142 West 36th Street
4th Floor
New York, New York 10018
Telephone: (212) 779-4972
Fax: (212) 779-6058
quartoknows.com
Visit our blogs at quartoknows.com

10 9 8 7 6 5 4 3 2 1

ISBN: 978-1-63106-030-4

Library of Congress Cataloging-in-Publication Data is available

Project Manager: Greg Oviatt
Technical Editor: Alexandra von Meier
Illustrations: Robert Steimle
Interior Design: High Tide Design
Cover Design: Heidi North

Printed in China

CONTENTS

Who today, can read a copy of The Inventions, Researches & Writings of Nikola Tesla, *published before the turn-of-the-century, without being fascinated by the beauty of the experiments described and struck with admiration for Tesla's extraordinary insight into the nature of the phenomena with which he was dealing? Who now can realize the difficulties he must have had to overcome in those early days? But one can imagine the inspirational effect of the book 40 years ago on a boy about to decide to study the electrical art. Its effect was both profound and decisive.*

—EDWIN ARMSTRONG, INVENTOR OF AM AND FM RADIO[1]

Introduction
Marc J. Seifer Ph.D.

The nineteenth-century philosopher John Stuart Mill is known for his "great person" theory. Do individuals or group effort determine history? It is Mill's contention that in many instances single individuals do indeed change the course of history. Isaac Newton, Galileo, Napoleon, Adolf Hitler, Winston Churchill, Walt Disney, Thomas Edison, Nelson Mandela, and Steve Jobs come to mind. Ultimately Chris Cooper's text *The Truth about Tesla* asks the question, had Tesla not lived what would our world have been like?

This simple question gets to the heart of the book but also to the process of invention itself, as well as Tesla's role in the development of the modern age. Cooper cites Supreme Court Justice Wiley B. Rutledge, who suggests that we should not use modern sensibility to try and understand the role someone played within their times. Rather, it makes more sense to see what contemporaries thought. And that certainly was my approach when I was trying to figure out for

myself what Tesla's role was in the development of so many different inventions.

In the case of the induction motor and Tesla's discovery of the rotating magnetic field, it is clear that Tesla was not the first to discover how to create a rotary effect by using two AC circuits out of phase with each other. Walter Baily preceded him and Galileo Ferraris came up with the same idea independent of Tesla's or Baily's work.

Two major factors to keep in mind in this instance are (1) What was the present-day technology before Tesla introduced his induction motor and AC polyphase system to the world in May of 1888? And (2) What did his contemporaries say about Tesla's invention?

This problem has to do with the very nature of alternating or AC current, which changes its direction of flow at many times per second. Trying to make it go in one direction involved the use of a commutator, which eliminated the reverse flow but at a great cost of efficiency. Thus all DC (direct current) or prevailing AC (alternating current) electrical transmission systems made use of

OPPOSITE: Tesla in his office in New York City, 1916

the commutator. Tesla's system did away with the commutator, and neither Baily nor Ferraris devised patents to achieve that same end.

Thus, before Tesla's May 1888 speech before the electrical community at Columbia College, the prevailing thought was that electricity, whether DC or AC, could only be transmitted about a mile, as power dropped off over distance, and only used for lighting. If a company wanted electrical power, it had to be situated along a waterfall or construct an enormous electrical coal-operated generator at the site of the plant.

Shortly after Tesla's talk, he sold his induction motor and 40-patent AC polyphase system to George Westinghouse for a king's ransom, and three years later, in the spring of 1891, Westinghouse transmitted electrical *power* 2.6 miles in Telluride, Colorado. That achievement was astonishing, but it was eclipsed the very same year in dramatic fashion when British engineer C. E. L. Brown of the Oerlikon company and Russian engineer Mikhail Dolivo-Dobrovolsky, from Allgemeine Elektrizitats Gesellschaft (AEG), transmitted electric power from Lauffen, Switzerland, to Frankfurt, Germany, a distance of 108 miles. Although Dolivo-Dobrovolsky tried to claim the invention as his own, Brown, a highly respected engineer and a man of integrity, said without equivocation that they succeeded by using Tesla's patents.[2]

Just six years later, Westinghouse put the system in at Niagara Falls, where through one single renewable source of clean energy, a waterfall, hydroelectric power came to electrify the entire northeast, covering a distance of 400 miles in any direction. Now millions of homes could be illuminated and run appliances, thousands of miles of streets could be lit at night, and factories run by electricity could be located anywhere. Although controversy still exists as to what Tesla's precise role was in this development, to a rousing applause, he spoke at the inauguration of the Niagara Falls power plant as the inventor, and his name is on the patent plaque for nine patents.

In citing a previous case on a similar issue, Judge Townsend responded to what today is called the "doctrine of obviousness" as such: "The apparent simplicity of a new device often leads an inexperienced person to think that it would have occurred to anyone familiar with the subject, but the decisive answer is that with dozens and perhaps hundreds of others laboring in the same field, it had never occurred to anyone before."[3] This is precisely the situation in Tesla's case.

Chris Cooper covers this topic from several angles and he also goes into the issue as to who invented wireless communication. One of the central themes of this book is that inventions are the product of many individuals. Certainly, Tesla did not harness Niagara Falls by himself. Westinghouse had an entire crew do that and, no doubt, hundreds of technical decisions had to be made by many highly intelligent engineers. Further, Tesla shared the patent plaque with three other inventors, Albert Schmid, Arthur Kennelly and William Stanley.

In the case of wireless, Tesla was certainly not the first to transmit electrical impulses without wires. As Cooper points out, a system very similar to Tesla's was developed by Mahlon Loomis twenty years earlier. And this system was not only patented, it was also funded by an act of Congress! However, there has to be a reason why Adolf Slaby, a highly regarded German engineer and contemporary of both Tesla and Marconi, called Tesla the "Father of the Wireless." And also why John Stone Stone, physicist and electrical engineer, educated at Columbia and also Johns Hopkins University, and president of the Institute of Radio Engineers, testified in 1916 that the present-day radio systems have developed to a technology that "returned to the state which Tesla developed it."[4]

Certainly radio developed because people like DeForest and Marconi spread this technology throughout the world, but the question remains, who was the inventor? Tesla's watershed lectures at Columbia College (now University) in 1891, before the Royal Societies in London and Paris in 1892 and at lecture halls before thousands in Philadelphia, St. Louis and at the Chicago World's Fair in 1893 were attended by the scientific elite, including Nobel prizewinners J. J. Thomson, Lord Rayleigh and Robert Millikan, and other luminaries such as Lord Kelvin, Andre Blondel, James Dewar, John Ambrose Fleming, Hermann von Helmholtz, Elisha Gray, Sir William Crookes, Oliver Lodge, Sir William Preece, Alexander Graham Bell and Elmer Sperry. Expanding on the work of such forerunners as Sir William Crookes and Heinrich Hertz, Tesla displayed precursor radio tubes and high frequency oscillators, which by means of wireless illuminated different cold lamps when different frequencies were reached. By setting up tuned circuits that responded when a combination of frequencies were transmitted, the seeds of so many groundbreaking creations were presented including the essential basis for fluorescent and neon lighting, the gyroscope, remote control, robotics, wireless communication, TV transmission, cell phone technology and radio guidance systems.

Did these developments happen all because of Tesla? That is the key question posed in the subtitle of this work: "*The Myth of the Lone Genius in the History of Innovation*." Chris Cooper will explain that rarely is there one inventor to any creation. And this text will explore not only Tesla's role in the history of these inventions, but also the role of a number of other key inventors in other groundbreaking creations including those of Alexander Graham Bell in the invention of the telephone and even the Wright Brothers in their invention of the aeroplane.

This is an exciting work that uses Tesla's life as a template to tell a larger story, one about how inventions are created and how history interprets the role individuals play in their development.

—Marc J. Seifer, Ph.D., author of *Wizard: The Life and Times of Nikola Tesla*

David Sarnoff conducts an Inspection Tour of RCA Transoceanic Station at New Brunswick NJ

1921

among Mr Sarnoff's guests are Albert Einstein, Steinmetz, Langmuir and other famous Scientists.

"It takes a thousand men to invent a telegraph, or a steam engine, or a phonograph, or a telephone or any other important thing—and the last man gets the credit and we forget the others. He added his little mite—that is all he did. These object lessons should teach us that ninety-nine parts of all things that proceed from the intellect are plagiarisms, pure and simple; and the lesson ought to make us modest. But nothing can do that."

—MARK TWAIN, LETTER TO HELEN KELLER, MARCH 17, 1903

chapter one

Myth of the Lone Genius

Every writer is an inventor. And like the best inventors, the best writers borrow heavily from others. Every fact you read here, every story recounted, has been told in existing biographical publications or can be found through a robust search of the Internet. No single component of this book is new, and yet the story of Nikola Tesla has never been told like this before. Like most ingenious ideas, the truth about Tesla has always been right there for anyone—from the most ardent fanatic to the most casual inquisitor—to find, if only he or she stumbles upon the right bits of information, in the right order.

And that is the point, really. Innovation, like history, is messy. It involves the passions and perplexities of people, with all their quirks and foibles. We like to think of innovation as occurring in leaps and bounds, society rapidly propelled forward by revolutionary concepts uncovered by the best and brightest among us. After all, human history is written in technological epochs, from the Bronze Age to the steam engine, the skillful harnessing of hot water to power an entire industrial revolution. This history is clean, if simplistic. It is nothing like the history we live, where technological progress is wrought in painstakingly small increments, as much from a clash of competing self-interests as from a wondrous sense of curiosity. It is a sanitized history, devoid of the kinds of daily struggles, for guts and for glory, that make life both messy and meaningful.

Patents and Prestige

On an unseasonably hot afternoon in June of 1943, the United States Supreme Court handed down its decision in *Marconi Wireless Telegraph Company of America v. United States*, the lawsuit that would put an end to the decades-old legal battle over just who invented the radio. Technically, the suit was the final attempt by

OPPOSITE: Tesla (9th from the left) with Albert Einstein (to his right), Charles Steinmetz (to his left) and other luminaries inspecting the New Brunswick Marconi Station

the Italian inventor Guglielmo Marconi and his American subsidiary to recover damages from U.S. companies for infringement of radio patents he had been issued in 1904. Since December 1901, when Marconi announced that he had successfully intercepted in Newfoundland a wireless signal transmitted from England, the inventor was heralded as the father of radio. Shortly thereafter, he quickly set about protecting his interests by applying for patents on both the method and apparatus he used.

Initially, the U.S. Patent Office rejected Marconi's application, citing (among other things) a patent issued in 1900 to Nikola Tesla for a wireless "System of Transmission of Electrical Energy" that the Serbian inventor claimed could be used "to transmit intelligible messages to great distances."[1] Although

Guglielmo Marconi

Marconi denied ever having read anything about Tesla's system, in October 1903 the U.S. examiner found Marconi's claim, well, patently absurd.[2] Only a year later, however, in a decision that remains a mystery to this day, the Patent Office suddenly reversed itself and granted Marconi patents for the most essential radio components.[3] To add insult to injury, the Nobel Committee granted Marconi—not Tesla—the Nobel Prize in Physics in 1909.

By 1915, Tesla had had enough and filed his own suit seeking to invalidate Marconi's patent.[4] He was, however, on the brink of bankruptcy, living off of loans from financiers whom he enticed by making extraordinary claims to world-changing inventions like electrical flying machines and the global transmission of wireless electricity. Tesla barely had money to feed himself, let alone pursue an expensive patent suit.

In any event, powerful financial interests aligned against Tesla, largely as a result of historical events few could have foreseen. By 1917, as the bloody events of World War I were just reaching their nadir, the U.S. government seized all patents owned by U.S.-based radio manufacturers. This allowed the military to use the technology for the war effort without being encumbered with pesky and expensive licensing fees. Although the war ended only a year later, the military sought to keep the national radio system it had created squarely in U.S. control. Navy admiral W. H. G. Bullard met with executives of the General Electric Corporation to strike a grand bargain. If GE agreed to stop selling its AM radio transmitters to the Marconi Company (and its U.S. subsidiary, the Marconi Wireless Telegraph Company of

America), the U.S. government would sanction a virtual monopoly on long-distance radio by an American-owned subsidiary controlled by GE. The executives agreed, quickly bought controlling interest in Marconi Wireless, and incorporated the Radio Corporation of America (RCA), after which the U.S. military granted the company rights to all of the radio terminals it had confiscated during the war. Thus, by the autumn of 1919, Marconi Wireless was in the hands of powerful industrialists who had every incentive to protect Marconi's patent.

In the meantime, Tesla had gone broke, he had suffered a nervous breakdown, his wireless patents had expired, and he simply did not have the wherewithal to pursue a lawsuit against these interests. Tesla's defeat, however, would not stop them from reinforcing Marconi's patent by bringing an infringement claim of their own in the U.S. Court of Claims. Twenty-seven years later, the case had reached the U.S. Supreme Court, which, in the swelter of early summer, wrestled with how to resolve the thorny issue once and for all.

The conventional wisdom is that Tesla was finally given due credit when Chief Justice Harlan Stone, writing for a five-member majority, declared that the inventor had, in fact, anticipated at least four major components of Marconi's radio system.[5] But the decision was hardly the resounding vindication many of Tesla's admirers claim. The inventor's role in the whole matter takes up fewer than four pages of an opinion that runs nearly sixty, most of which are devoted to reviewing how the essential components of Marconi's system had borrowed heavily from inventors other than Tesla—namely, the American physicist John Stone

Justice Felix Frankfurter

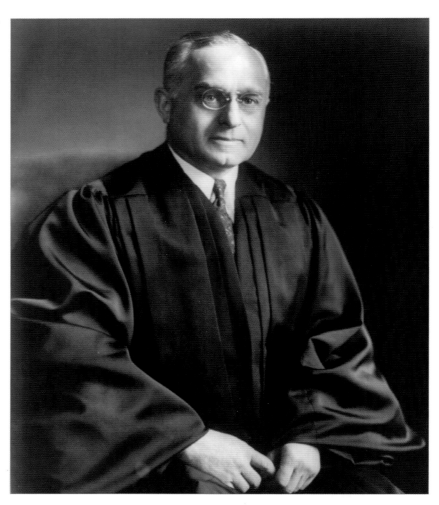

Stone and two British inventors, Sirs John Ambrose Fleming and Oliver Joseph Lodge.

After grappling with the issue for weeks and reading over a draft of the majority opinion, Justice Felix Frankfurter refused to side with his colleagues.* Consistent with his penchant for judicial restraint, Frankfurter penned a florid dissent, cautioning the court from using hindsight to recreate how scientific innovations occur:

---

* Justice Frankfurter was joined in his dissent by Justice Owen Roberts. Ultimately, the Court's opinion was supported by five justices, opposed in part by three, with one justice recused.

The discoveries of science are the discoveries of the laws of nature, and like nature they do not go by leaps. Even Newton and Einstein, Harvey and Darwin, built on the past and on their predecessors. Seldom indeed has a great discoverer or inventor wandered lonely as a cloud. Great inventions have always been parts of an evolution, the culmination at a particular moment of an antecedent process. So true is this that the history of thought records striking coincidental discoveries, showing that the insight first declared to the world by a particular individual was "in the

innovation is the product of ingenious minds working in relative solitude, slaving away at intractable problems, motivated by the promise of exclusive rights to whatever world-changing epiphany suddenly hits. A fundamental assumption of U.S. patent law is that inventors would not be bothered to chase innovations if not for a guaranteed period of total control over the economic fruits of their labor. Nowhere is this assumption more obvious than in the words of Abraham Lincoln, which are etched above the north entrance of the original Patent Office building: "*The patent system adds the fuel of interest to the fire of genius.*"[7]

More recently, the issue of whether the U.S. system of patent protection encourages or discourages innovation has been the subject of much debate. On that question, the jury appears to be out. Certainly a plethora of sociological studies have found that patents, by holding out the prospect of substantial economic reward, do induce invention.[8] In countries without patent protections, for example, several studies have shown that people have an incentive to simply copy others rather than invent new, more efficient methods themselves.[9] A 2008 study by the Organization for Economic Cooperation and Development (OECD) concluded that strong patent protections contribute to higher research and

air" and ripe for discovery and disclosure. The real question is how significant a jump is the new declaration from the old knowledge.[6]

Frankfurter went on to express that he had little doubt that the great transforming forces of technology that were shaping mid twentieth-century America had rendered obsolete much of U.S. patent law. Now, more than seventy years after that Supreme Court decision, much of our patent law still remains enamored of the myth that

development (R&D) expenditures, particularly in high-tech industries.[10] The assumption is that R&D expenditures, at the very least, reflect effort at invention.

The assumptions behind the modern patent system have come under attack, however, by sociologists and economists, who argue that studies supporting the theory that patents spur innovation are either flawed or denied by more empirical evidence proving just the opposite. There is a growing body of academic literature demonstrating that patent protections discourage innovation by stifling collaboration. When researchers from M.I.T. and the Harvard Business School, for example, tested whether loss of profits reduces the incentive to invest in R&D, the results were surprising. Rather than jealously guard incremental developments, modern inventors (both individuals and firms) were, according to the findings, more likely to voluntarily relinquish exclusive intellectual property rights in exchange for the additional innovations that resulted from freely sharing information in an open-access platform.[11] It appears that cutting-edge innovators—even those with a strong profit motive—increasingly recognize that there is more value to be had from developing the best ideas cooperatively than in exercising exclusive control over a merely good one. Other researchers set up an interactive game simulating the behavior of competitive inventors operating with and without patent systems. They found that a system with patent protections generates significantly lower rates of innovation than does an open-access system, where inventors freely share ideas.[12]

For the patent naysayers, there are several reasons that collaboration is more likely to spur innovation (and do so more quickly) than competition. First, modern innovations generally involve what social psychologist Dr. Dean Keith Simonton calls "happy combinations," chance associations between ideas, stumbled into more or less blindly by subject experts.[13] The more people with access to information, the more happy combinations will be found by people with different perspectives finding different associations. Moreover, if good inventions are the result of new combinations forming better products, then collaboration will more quickly validate meritorious ideas and discard less useful ones, speeding the iterative process required to produce world-changing innovations.[14]

The Genius Myth

While this book is not likely to settle the debate over whether patent protections spur innovation or stifle collaboration, it does seek to explore how mainstream history and conventional wisdom, like U.S. patent law, fabricate myths about the process of invention and construct cults of personality around certain inventors in a way that ultimately obscures the true nature of innovation. In a 2014 op-ed for the *New York Times*, innovation guru Joshua Wolf Shenk declared the end of "genius," at least as a way of describing individuals.[15] Shenk is among a new group of sociologists seeking to dispel the long-accepted notion that invention is the product of solitary creators, lone geniuses striving against the tide of history to generate a truly singular idea. Despite a considerable amount of evidence that creativity results from social interactions, the conventional wisdom continues to be that creative genius cannot be learned. One is either blessed with it or not. In the popular mythology of creativity, the social environment is, if anything, a hindrance to the creative process.[16] Psychologists Alfonso Montuori and Ronald Purser described this idea as the myth of the lone genius:

> *This modern view of creativity has venerated the artist or genius as cultural hero, because he or she is someone who has forged something new and original by struggling against and rising above limiting, stultifying forces of the conforming masses. To maintain such a stance, the creative person must in a sense disengage him- or herself from the environment. The resulting psychic isolation, along with what are perceived to be the "deviant" "schizoid" behaviors of the creative person, is romanticized or even seen as being synonymous with genius.*[17]

Modern media—from books and magazines, to film and television—seduce us with the simplicity of this myth. The story of a singular great mind heroically overcoming the collective economic and social forces aligned against him makes for compelling entertainment. But it obscures the complexity of innovation and does a disservice to the thousands of individuals whose countless contributions to social progress are known only to the most avid historians. The myth also turns history from a collective endeavor into a game of winners and losers, using such imprecise and ignoble devices as patent law to determine who will be placed in the pantheon of human greatness. If legacy is the only

true immortality, then this version of history amounts to canonization by cockfight.

The myth of the lone genius has been greatly perpetuated by the recent historical redemption of Nikola Tesla, the Serbian inventor who, we are told, invented everything from radar to the microwave oven, but whose magnanimity and epic battle with Thomas Edison left him destitute, yammering at pigeons and relegated to the dustbin of history. It has become something of a fad to restore Tesla to his rightful place among the persecuted geniuses who were simply ahead of their time. Tesla fanaticism, in fact, has reached a fever pitch in recent years driven (metaphorically, if not actually) by an electric car, countless documentaries, endless blog posts, and a high-profile effort to construct a museum on the Long Island property where the inventor tried to transmit free, wireless power around the globe. It is reflected in recent biographies declaring the electrical engineer a "Wizard," "Master of Lightning," and "The Man Who Invented the Twentieth Century."

In dispelling the myth behind the man, this book is not intended to disparage Tesla or unseat him from his rightful place in history as one of the world's most visionary electrical innovators. Rather, by unmasking the mythology of Nikola Tesla, the pages that follow uncover how the myth of the lone genius has come to shape the history of innovation, even as told by the most meticulous of scholars. Tesla's story helps reveal how cults of personality are crafted in the zeal to venerate those whom history may have forgotten. But, this zealotry often runs the risk of repeating the mistakes it seeks to redress. Prudence gives way to passion, and even those attempting to correct history unwittingly contribute to its distortion.

Tesla wireless article from the *Washington Herald*, March 17, 1912

THE TRUTH ABOUT TESLA

For this object lesson, Tesla is practically ideal. For most of his biographers, his vindication is tied to patent case victories. Tesla is credited with inventing the polyphase alternating current motor because he won the patent battle. Ditto with radio (in the end). Ironically, these same historians deride the patent system when it does not comport with the simplistic lone genius myth. When Tesla loses a patent battle, it is because powerful interests have thwarted the legal system, at least for a time. But if powerful interests can prevent the courts from recognizing Tesla's genius, can't they also compel the courts to falsely proclaim it?

History is written by the (patent case) winners, and so it is for Nikola Tesla. But is it written accurately? Unwittingly or not, many of Tesla's biographers marry prestige to patents, often forgetting that patent law—especially in the U.S.—assumes a process of invention often out of touch with reality. Invention is rarely a series of discrete ideas, conceived in isolation.[18] Rather, it is an incremental process, involving sparks of insight whose importance may be unknown or unappreciated, even by those who have had them.

The pages that follow reveal the truth about Tesla (and, in the process, the true nature of innovation) by piecing together a more accurate story of three of the most significant inventions attributed to him—the alternating current (AC) motor, the Tesla coil, and wireless transmission of electricity. The beguiling sagas surrounding how Tesla came to be credited with these inventions may raise more questions than they answer. History, like innovation, is messy. But in the process of telling a more accurate (yet no less entertaining) history of the inventor, we can better uncover a more accurate (yet messier) process of invention.

"*Let the future tell the truth, and evaluate each one according to his work and accomplishments. The present is theirs; the future, for which I have really worked, is mine.*"

—NIKOLA TESLA, INTERVIEW IN *POLITIKA*, 1927

chapter two

The Conventional Account of a Modern Genius

The conventional account of Nikola Tesla's life reads like a daytime drama, full of mystery and tragedy, and with more than a touch of embellishment. Most of what we know about his early life—his childhood, upbringing, and most of his professional endeavors prior to being crowned the "Electric Messiah" and father of the modern age—we learn from Tesla himself. Most, in fact, derive from brief accounts revealed in the slim volume *My Inventions*, originally published in 1919 as a series of articles in the journal *Electrical Experimenter*. With unrestrained flare, Tesla gives what he calls a "faithful account" of his life and his eventual realization that he was, indeed, an inventor (or as he modestly puts it, "one of that exceptionally privileged class without whom the race would have long ago perished in the bitter struggle against pitiless elements").[1]

Even the account of Tesla's birth is shrouded in a kind of mystery and drama that fits nicely

OPPOSITE: Tesla in 1879 at age 23 RIGHT: *The Electrical Experimenter* Tesla autobiographical series

NOVEMBER, 1916   15 CENTS

The Electrical Experimenter

POPULAR   ELECTRICAL   NEWS   ILLUSTRATED

LIGHTNING MADE TO ORDER SEE PAGE 474

★   LARGEST CIRCULATION OF ANY ELECTRICAL PUBLICATION

with the quintessentially American myth of the independent hero facing (and overcoming) overwhelming odds. Tesla was born, as the legend goes, in 1856 at the stroke of midnight between July 9th and July 10th. As a violent electrical storm raged overhead and little Nikola was fighting his way out of the womb, a frightened midwife was said to have declared that the newborn would be a "child of the storm," to which Tesla's mother replied, "No . . . of light!"[2]

Whether or not the universe was heralding the entrance of a rare genius has been lost with the weather records of rural Croatia. But, Tesla—a man who would wander to (and around) America, never bear children, and die alone in a New York hotel room—was born a Cancer, a sign said to be maternal, domestic, and sedentary, with large families. Either the universe has an ironic sense of humor or, more likely, was far less involved than the conventional account of Tesla's birth might suggest.

Childhood: Genius by Instinct

If you troll the Internet long enough, you will find factions variously describing Tesla as Serbian, Croatian, or even Romanian. Belgrade has both a museum and an airport dedicated to the inventor. Not to be outdone, Croatia opened a museum and theme park in Tesla's hometown on the 150th anniversary of his birth. In addition, an entire online forum is devoted to uncovering Tesla's Romanian heritage largely by tracing the alleged etymology of his family name.

While Tesla was ethnically Serbian, he was born in Smiljan in a province in what is today Croatia. Given the messy history of the Balkans—a history involving among other things the collapse of the Ottoman Empire and the impact of the Napoleonic wars—who gets to claim official credit for the inventor is murkier than the Black Sea. During the late fifteenth and early sixteenth centuries, the Ottoman Turks controlled vast areas of the Balkan Peninsula as far west as modern-day Kosovo. The largely Christian Serbs they displaced did not flee far. Under the encouragement of the Catholic Hapsburgs (who would rule the Austro-Hungarian Empire), most of these Serbs settled into a kind of militaristic buffer zone that the Croatian Parliament organized just outside the Turkish-controlled areas of their country. Tesla's ancestors were among these displaced Serbs, migrating from western Serbia to the Croatian province of Lika as early as the 1690s.

Tesla was named for his paternal grandfather, Nikola, who was born in Lika in 1789, shortly before the Hapsburgs ceded that part of Croatia to Napoleon. Truth be told, Napoleonic control may have contributed more to Tesla's appreciation of science and mathematics than either his Serbian ethnicity or his Croatian nationality ever did. The French swept out ancient monarchies and long-held superstitions and encouraged a new enlightenment by building schools and colleges to introduce science and rationality to the underclasses.[3] It was during this period of French influence that Tesla's grandfather married Ana Kalinić, the daughter of a colonel in Napoleon's army. She bore him two sons, Josif, Tesla's uncle (who would help bankroll Tesla's secondary education), and Milutin, Tesla's father. While Josif would follow in his father's hawkish footsteps and become a professor of mathematics

at the Austrian Military Academy, Milutin chose more dovish pursuits and became a priest in the Serbian Orthodox Church.

In 1847, Milutin married Djuka Mandić, the daughter of an Orthodox priest from Gračac, a town along the southern border of Lika. Djuka's grandfather, all of her uncles, and several of her brothers were all priests. From birth, therefore, Tesla was expected to follow the family tradition and enter the clergy. But, even in childhood, young Nikola's ambitions tended more toward the scientific than the spiritual.

When he was sixty-two years old, Tesla recounted his life, writing of a few bizarre events that would give his early years a patina of the supernatural, if not the downright incredible. First, Tesla described having suffered from "a peculiar affliction" during his childhood, including "the appearance of images, often accompanied by strong flashes of light," that neither psychology nor physiology could explain. Although most of us would describe these afflictions as the mere daydreams of an active imagination (or the hallucinations of a disturbed mind),* Tesla assured his readers that they were not, since, in all other respects, he was normal:

> When a word was spoken to me the image of the object it designated would present itself vividly to my vision and sometimes I was quite unable to distinguish whether what I saw was tangible or not. This caused me great discomfort and anxiety . . . To give an idea of my distress, suppose that I had witnessed a funeral or some such nerve-racking spectacle. Then, inevitably, in the stillness of the night, a vivid picture of the scene would thrust itself before my eyes and persist, despite all my efforts to banish it. Sometimes it would remain fixed in space though I pushed my hand through it.[4]

Although Tesla claims that these images were always "pictures of things and scenes which I had really seen, never those imagined," later

---

* The illustrious biographer Margaret Cheney even suggested that, had Tesla been taken to a psychologist as a child, he most likely would have been diagnosed as schizophrenic and prescribed drugs to "'cure' the very fountain of his creativity." See Cheney, *Tesla: Man Out of Time* (New York: Simon & Schuster: 1981), 35.

he would claim to harness this affliction to conjure detailed images of inventions never seen (nor even put to paper) and run whole experiments as if he had built and tested the contraptions in a laboratory.

This remarkable (if bizarre) ability to control hallucinations and use them to work through complex engineering experiments would later be seen as proof that Tesla possessed almost supernatural mental acuity, something that would set him apart as a lone genius among mere mortals. Indeed, in an exhaustive 2013 biography of Tesla, W. Bernard Carlson, professor of science, technology, and society at the University of Virginia's School of Engineering and Applied Science, described Tesla's imagination in exceptional terms, as "unusually powerful" and the singular aspect of his childhood that would prove most essential to his success as an inventor.[5]

A second incident that Tesla recalled from his childhood reinforces the image of the inventor as endowed with a kind of scientific intuition few mortals can achieve. Distraught over the sudden death of Tesla's older brother, Dane (who perished after being thrown from the family's prized Arabian horse), Milutin moved the family from Smiljan to the larger village of Gospić, the administrative seat of the province. There, at age ten, young Nikola enrolled in what today we would call junior high school. Although he excelled at mathematics, he nearly flunked out when he could not pass the required drawing class. Milutin, the popular cleric of Gospić's Church of the Great Martyr George, was forced to intervene with school officials so that Nikola was not thrown out of school altogether. Curiously,

in Carlson's account of Tesla's childhood, he does not depict Tesla's inability to translate his stark visualizations to paper as evidence of the young inventor's artistic mediocrity. Rather, he presents Tesla's difficulties as further evidence of an exceptional nature, attributing Tesla's failures to the inventor's "preference for undisturbed thought."[6]

Despite his academic shortcomings, Tesla apparently won widespread acclaim. Shortly after moving to Gospić, the town received a new fire engine and, under the leadership of a local merchant, organized a uniformed fire department. After parading the new red-and-black pump through the streets, a team of sixteen firemen set about demonstrating the engine by feverishly pumping the contraption's handles up and down—but no water came out. While the whole town stood perplexed, young Nikola suddenly waded into the river where the engine's input hose had become blocked by a kink. He located the obstruction, straightened the hose, and water immediately began spurting from the other end.

According to Tesla, for this relatively unremarkable feat, the grateful townspeople hoisted him on their backs like an Olympian, praising him as the hero of the day. It is probable that Tesla had been exposed to basic engineering concepts, either by observing the many household tools his mother improvised or by reading any number of his father's growing library of scientific books. Nevertheless, Tesla reminded readers that his knowledge of the fire engine's mechanics "was nil and I knew next to nothing of air pressure, but *instinctively* I felt for the suction hose in the water and found that it had collapsed . . . Archimedes running naked through the streets of Syracuse and shouting *Eureka*! at the top of his voice did not make a greater impression than I did that day."[7]

By Tesla's own accounts he was a sickly child, falling prey to a host of illnesses during his teens, including malaria, which he contracted in the marshy lowlands of Karlovac (where he traveled to attend the Croatian equivalent of high school) and cholera, which left him bed-ridden for nine months (upon his return home). The latter sickness was so grave that a coffin was ordered and Tesla's distraught father called to the boy's bedside. Seeing an opportunity, Tesla seized on this near-death experience to suggest to his desperate father, "perhaps I may get well if you will let me study engineering." Milutin, either grasping for any hope of the boy's recovery or assuming Nikola was so far gone that he would never have to make good on his promise, let go of his own hopes that Tesla would follow him into the clergy and told his son he would send him to the best technical institution in the world.

Remarkably, Tesla soon recovered. Unfortunately, however, he had reached the age when all Serbian boys were required to give three years of compulsory service in the Austro-Hungarian army. But Milutin, fearing that Nikola would never survive the strenuous duties of a soldier, thought it better that the boy flee to the mountains outside of Gospić. So Tesla wandered the countryside of Croatia from the fall of 1874 until the summer of 1875, carrying only a hunting rifle and a bundle of books.[8] Despite his son's frailty, Milutin somehow managed to secure a scholarship from the Military Frontier Administration Authority for Nikola to attend the Joanneum Polytechnic School in Graz, Austria, in exchange for eight years of military service upon his graduation.[9] Perhaps Tesla's father assumed Nikola would toughen with age. Or perhaps he merely felt obliged to honor the promise he made to his son. In any event, off to Graz Tesla went, in the hope of becoming an engineer and the expectation of becoming a man.

University: The Prodigal Son Comes of Age

When Tesla first arrived at the Polytechnic School, he focused his studies on mathematics and physics, perhaps with the intention of following in the footsteps of his uncle, Josif, and becoming a professor. While the school was renowned for its civil engineering curriculum, it had no official course of study in electrical engineering. Most of what Tesla learned of electricity came from the physics classes of Professor Jacob Poeschl, a "methodical and thoroughly-grounded German," a bear of a man whom Tesla described as having "enormous feet and hands," but whose classroom experiments were performed with clocklike precision.[10]

By every account, during his first year in Graz, Tesla worked tirelessly, with an almost obsessive diligence. He recounted how his compulsion to finish reading the complete works of Voltaire (almost one hundred volumes of small print) nearly killed him. Nevertheless, by the end of his first year he successfully completed nine final exams, more in a single year than any former student. He claimed to work from 3:00 A.M. to 11:00 P.M. every day, "no Sundays or holidays excepted." Apparently he managed to maintain this regimen by consuming copious amounts of coffee. (In his memoirs,

Tesla even acknowledged that "the truth is that we need stimulants to do our best work under present living conditions.")[11] Returning home with top grades, Tesla was crestfallen when his father reacted ambivalently. Later, Tesla learned that several of the Polytechnic professors had written his father warning that his son was likely to kill himself with overwork. In fact, Tesla had started to experience heart palpitations from all the caffeine he was consuming and had to limit his coffee consumption, a moderation he would practice his whole life.

When Tesla returned to Graz for his sophomore year, he learned that the Military Frontier Administration was being abolished and he would be losing his fellowship.[12] Since his father would be unable to afford the steep tuition on a priest's salary, Tesla knew it was likely he would have to drop out before the end of the school year.

It was during one of Professor Poeschl's lectures that sophomore year when Tesla is said to have begun his quest to perfect the electrical motor. The school had obtained a Gramme dynamo, one of the first practical direct current (DC) generators.† Though popular at the time, Gramme dynamos had one significant drawback: they created dangerous sparks because they needed to be fit with a commutator, a

---

† Prior to this design by Belgian engineer Zénobe-Théophile Gramme, even dynamos with commutators would produce unstable currents, constantly building a charge, discharging, and then dropping nearly to zero voltage again.

Joanneum Polytechnic School (Graz, Austria)

clunky mechanical mechanism that helps create a smoother, direct current (see p. 90). While watching Poeschl deal with these sparks during one memorable demonstration, Tesla suggested that it might be possible to create a generator that didn't require the commutator brushes at all. Though fond of Tesla, Poeschl decided to use his prize student's ruminations as a "teachable moment," launching into a lecture on electrical engineering and declaring that Tesla's imagined generator "would be the equivalent to converting a steadily pulling force, like that of gravity, into a rotary effort . . . a perpetual motion machine, an impossible idea."[13]

Though silenced by Poeschl's rebuke, Tesla remained committed to the idea of a sparkless generator. Nothing Poeschl said would deter

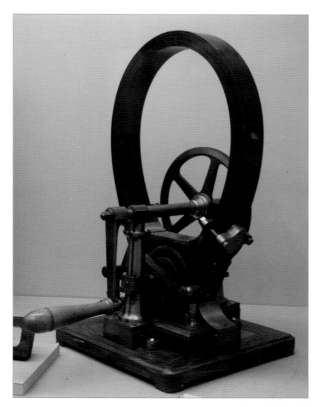

Gamme Dynamo Machine

Tesla from searching for a solution. Looking back, he would attribute his perseverance to "something which transcends knowledge," an innate ability to perceive truths "when logical deduction, or any other willful effort of the brain, is futile."[14] "The young Serb had no idea how it might be done," admitted biographer Margaret Cheney in her seminal account of Tesla's life, "but *instinct* told him that the answer already lay somewhere in his mind."[15] According to Tesla's account, if hindsight is 20/20, an inventor's instinct is downright prophetic. Knowledge can take a siesta, while the inventor waits for that Archimedes-like *Eureka!* moment when the answer presents itself, presumably out of thin air.

His money running out, Tesla took up gambling—in particular billiards and card games—to try to support himself. For a mathematical genius, it appears Tesla was never that good at poker, though he was a natural with a pool cue, becoming "almost professionally skillful" at the game.[16] Still, in his third year in Graz, gambling became less a means to an end and more an obsession. Although by Tesla's account, he spent his remaining time in Graz fruitlessly trying to perfect his brushless DC generator, the historical record indicates he spent most (if not all) of it gambling and carousing. He stopped going to classes entirely. In fact, the Polytechnic School has no record of him even registering for classes for the spring semester of his junior year (1878).[17]

Flat broke—and without telling a soul—in late 1878, Tesla left Graz for Maribor (a small city in modern-day Slovenia). His friends feared he had become depressed and drowned himself in Graz's Mur River. His family knew only that he had disappeared. By sheer chance, one

Pool Prodigy or Clever Hustler?

Tesla biographer Marc Seifer noted that Tesla could be deceptive, particularly in pecuniary matters. But the inventor also was not above deceit when the only matter was perception of his admittedly extraordinary abilities, especially his proficiency at billiards. He spent the better part of the latter half of his sophomore year in Graz drinking and playing pool at the Botanical Gardens. By all reports Tesla became an accomplished pool player, a skill he used to endear himself to the few Americans employed in Edison's Paris operation.

When the inventor arrived in America, however, it appears he purposefully obscured his skills. In his memoirs, Alfred O. Tate, Edison's personal assistant, apparently unaware of Tesla's background in billiards observed that the inventor "played a beautiful game." Tate wrote of Tesla's short employment at the Edison Machine Works, when the young inventor stopped working only for the rare nap and the occasional game: "He was not a high scorer, but his cushion shots displayed a skill equal to that of a professional exponent of this art."

On more than one occasion, Tesla entertained the false notion that he derived his skill at billiards merely by watching others play, and he was not above using the ruse to hustle money from his opponents. Lorenzo Delmonico, proprietor of the famous New York City restaurant, recalled that patrons "managed to make him play pool one night." Noting that the inventor "had never played," but appeared to "study out pool in his head," Delmonico told one reporter, Tesla then "beat us all and got all the money…after we had practiced for years!"

of Nikola's former college roommates ran into him in Maribor and alerted Tesla's family of his whereabouts. Milutin immediately chased his son down and pleaded with him either to return home or to resume his studies in Prague. Tesla refused. Milutin returned to Gospić heartbroken and died a month later. In the meantime, Tesla was arrested in Maribor as a vagrant and deported back to Croatia.

Shamed by the impression he had left on his deceased father (and after drowning his sorrows in a months-long beer-fueled gambling bender), Tesla resolved to do as Milutin had suggested and finish his studies. Ever the prodigal son, he convinced his maternal uncles, Petar and Pavle, to lend him the money to complete his education, and in 1880, he set out for Prague. Though he claims to have spent two years in the Bohemian capital completing his degree, the Czech government can find no record that he was officially enrolled in any of Prague's universities during the time Tesla claims to have resided there.[18]

Eventually, Tesla's uncles became suspicious and stopped financing him. In January 1881, when frugal living threatened to become complete destitution, Tesla decided to move to Budapest. He had read in a newspaper that Ferenc Puskás, the brother of a Transylvanian entrepreneur (whom his uncle Pavle had served with in the Austro-Hungarian cavalry), had been authorized by Thomas Edison to supervise the construction of a telephone exchange in the Hungarian capital. Tesla figured he could use his family connections and technical skills to land a job working for Puskás. But by the time Tesla arrived in Budapest, Ferenc had not yet secured the financing for the enterprise and Tesla found himself begging for a position as a draftsman (despite his lackluster drawing skills) in the Central Telegraph Office of the Hungarian Government. Clearly dissatisfied with the work, Tesla quit the job after a few months, had a nervous breakdown, fell into a deep depression, and sequestered himself in the tiny room he had rented.[19]

Worried that Tesla might die of melancholy, his best friend, Anital "Tony" Szigeti (whom Tesla describes as having the body of Apollo, from the neck down), resolved to drag Tesla out of bed and force him to get some fresh air. Szigeti convinced the anguished inventor to walk with him each evening through the city park. It was during one of these walks, while quoting Goethe's *Faust* from memory, that Tesla claims he had his *Eureka!* moment and his design for a brushless polyphase AC generator came to him "like a flash of lightning."[20]

Soon after Tesla's epiphany, Puskás secured financing for the Budapest telephone exchange and was finally able to hire Tesla. The young inventor threw himself into the work,

impressing Puskás so much that the entrepreneur invited him‡ to Paris to work for Edison's French subsidiary, which had been hired to construct an incandescent lighting system to illuminate the city.

At Graz, Tesla had learned the theory behind electrical induction. But it was in Paris that he received his first practical education in electrical engineering. It was also where he first proposed to his supervisors his idea of producing a rotating magnetic field that would generate three separate alternating currents delivered over six separate copper wires. To Tesla's disappointment, the men who ran Edison's Paris outfit showed little interest in the young Serb's musings.

Although conventional accounts imply that Edison's associates either were too fascinated with DC technology or too myopic to jump at Tesla's innovation, there was a more

---

‡   And, incidentally, Szigety as well. Though a master mechanic, no evidence indicates that Tesla's athletic companion had more than passing experience with electrical engineering.

Vajdahnyad Castle in Budapest's City Park, where Tesla and Szigeti often walked

Tesla's Men: Speculations on the Inventor's Sexuality

When Tesla was in his sixties, he struck up a close friendship with a nineteen-year-old science journalist named Kenneth Swezey. Clearly enamored of the inventor, Swezey wrote that Tesla often would greet him at the door stark naked.[21]

Most biographers attempt to explain Tesla's penchant for male company and lifelong celibacy as a kind of austerity required of ingenious invention.[22] Carlson, at least, acknowledged that Tesla undoubtedly was attracted to men.[23] But even during his lifetime, rumors of Tesla's homosexuality abounded. According to former members of the American Institute of Electrical Engineers (AIEE), fear that stories about Tesla's sexual episodes (particularly his voyeurism) might become public ultimately prevented the inventor from being elected the Institute's president.[24]

By Tesla's own account, he "never touched a woman."[25] But for a time he maintained a separate Park Avenue apartment in the swanky Hotel Marguery in which (he confided to Swezey) he met and entertained "special" friends and acquaintances.[26]

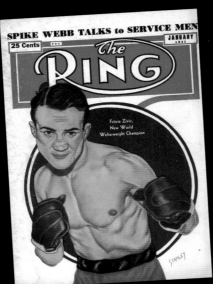

He clearly preferred athletic men, often inviting to the apartment boxers like the svelt Henry Doherty or the Yugoslav welterweight champion Fritzie Zivic.[27] Throughout his life he employed assistants

Fritzie Zivic

in their early twenties and did not shy from commenting on their physiques.

Tesla described Anthony Szigeti, the intimate friend with whom he took sunset strolls in Budapest, as having "the body of Apollo."[28] The two were inseparable, and Tesla found a way for the young Hungarian to follow him first to Paris, then to America. When he left Tesla (or died) under mysterious circumstances in 1891, Tesla confessed that, "I would have much desired to see him, because I would have wanted him."[29]

In Colorado, Tesla spent nearly every evening in his hotel room with his twenty-five-year-old assistant Fritz Lowenstein, who had emigrated from Czechoslovakia. However, the young engineer left Colorado in September of 1899 after Tesla and he had a falling out over letters the inventor discovered from Lowenstein's fiancée.[30]

In his mid-thirties, Tesla enjoyed an intimate friendship with the dashing twenty-eight-year-old Lieutenant Richard Hobson, a hero of the Spanish-American War. The two were introduced by Robert Underwood Johnson, who noted cryptically in his correspondence that the two "of course have much in common."[31] They enjoyed a playful correspondence and flirtatious friendship that neither appeared to hide. Tesla even signed one holiday greeting to Johnson's daughter "Nikola Hobson," though he was known to jealously guard the handsome lieutenant from his closest female confidant, Johnson's wife Katherine.[32] When Hobson decided to marry in 1905, Tesla was surely disappointed, despite the lieutenant's claim that the inventor occupied "one of the deepest chambers in my heart," and despite his insistence that Tesla serve as an usher at the wedding (since Hobson wished "to feel [Tesla] present standing close to me" on the occasion).[33] Even after the wedding, however (and well into Tesla's fifties), the two would meet at least once a month, ostensibly to go to the movies and then talk for hours. According to Hobson's wife, often her husband would not return from these sojourns until well past midnight.[34]

practical explanation for their indifference. Copper was often the single largest cost in an electrical installation and Tesla's system would have used nearly twice as much as the three-wire DC system Edison was promoting in the 1880s.[35] One of Edison's superintendents, David Cunningham (a minor figure, likely incapable of raising significant capital), however, was intrigued enough to offer to form a stock company with Tesla. But the budding genius declined, admitting later that he "did not have the faintest conception of what [Cunningham] meant."[36]

America: The Brave Man in the New World

In the spring of 1884, Charles Batchelor, Edison's right-hand man in France, was tapped to manage the Edison Machine Works in New York. Having taken a liking to the talented (if eccentric) Serb, Batchelor invited Tesla to join him and try his hand at improving the dynamos Edison's company was producing in New York. Tesla liquidated what few assets he had, borrowed additional funds from his uncles, and set sail for New York aboard the steamer SS *City of Richmond*. The trip was, by Tesla's account, an adventure

LEFT: Charles Batchelor

BELOW: SS *City of Richmond*

worthy of its own book. Allegedly, most of his money and belongings were stolen,§ the passengers attempted an unsuccessful mutiny, and Tesla was nearly thrown overboard in the ruckus.[37]

Conventional wisdom is that Tesla arrived at Edison's New York offices with no more than $2 in his pocket and a letter from Batchelor to Edison stating, "I know two great men and you are one of them; the other is this young man."[38] However, Carlson notes that Batchelor was already in New York (with Edison) before Tesla even set sail. Moreover, in a letter Batchelor

penned to Edison at least a year after Tesla started working for him, he referenced three European employees that he could mention as "capable as far as their work shows," none of whom was Nikola Tesla. And "while there are others capable," Batchelor admitted, "I think these are the best."[39] If Tesla had made the kind of impression upon Batchelor that the letter of introduction implies, surely he would have mentioned him when listing his best employees. More likely, the letter was from Tivadar Puskás—Ferenc's brother and business partner—who had made quite an impression himself when he arrived at Edison's Menlo Park laboratories in 1877 flashing a roll of thousand dollar bills in an (ultimately successful) attempt to convince the inventor to let the Puskás family franchise Edison's telephone and phonograph services

---

§   Interestingly, Cheney records a completely different account in which Tesla was just about to board the train from Paris to the coast when he discovered that most of his money—as well as his train and ship ticket—was missing. Tesla apparently scraped up enough change for another train ticket and eventually talked his way aboard the passenger ship *Saturnia*. See Cheney, *Tesla*, 48. The account presented here is far more likely since, by the time of Tesla's passage, the *Saturnia* had been acquired by Italia di Navigazione, S.p.A., which generally ferried passengers between Italy and New York. The *City of Richmond*, on the other hand, was used almost exclusively on the Inman Line between Paris and New York.

Castle Garden Immigration Center

in Europe.[40] Though the conventional account makes for great storytelling, Tesla likely never needed Puskás's letter anyway, since it is probable the two had already met when Edison had visited his Paris operations in the early 1880s, during the time Tesla was working there.[41]

In any event, Tesla arrived in New York on Friday, June 6, 1884. By varying accounts, including Tesla's own shifting recollections,¶ after making his way through the immigrant inspection station at the Castle Garden Immigration Center (which incorrectly recorded him as hailing from Sweden), he managed to ask a police officer for directions to Edison's offices in downtown Manhattan, pass a local machine shop where a mechanic was struggling with an electric motor, stop and fix the motor, collect $20 (equal to almost $500 today) for his services, find his way (on foot) to Edison's offices, volunteer for a work crew being dispatched that very moment to repair two ailing dynamos aboard the SS *Oregon* (a ship that held the record for the fastest transatlantic passenger run), *and* work through the night successfully repairing and testing generators so that the *Oregon* could depart on the morning of Saturday, June 7th (setting, incidentally, a new record for an eastbound transatlantic run).

While returning to the offices around 5 A.M. Saturday morning, Tesla claims to have run into Edison and Batchelor on the street.[42] According to Tesla, Edison remarked, "Here is

¶ According to Marc J. Seifer, an acknowledged authority on Tesla's life, the inventor's accounts of his first days in America "differ markedly depending on his mood at the time and his awareness of the size and shape of the audience." See Seifer, *Wizard: The Life and Times of Nikola Tesla: Biography of a Genius* (New York: Citadel, 1998), 34.

SS *Oregon*

our Parisian running around at night."[43] Tesla explained that he had just come from repairing the *Oregon*. Meeting this announcement with stunned silence, Edison walked away muttering to his companion, "Batchelor, this is a d—n good man."[44] And on the next day (a Sunday), while God rested, Tesla began his first day as an official employee of the Edison Machine Works.[45] It is a remarkable—if ultimately unbelievable—first forty-eight hours in the New World!

Tesla claims that for almost a year he worked tirelessly, from 10:30 A.M. until 5 A.M. the subsequent morning. But official company records indicate that he worked for the Edison Machine Works for only six months, a fact he verified himself in testimony during a later patent dispute.[46] According to Tesla, one of

ABOVE: Thomas A. Edison, with Batchelor and Tesla

RIGHT: Thomas A. Edison

Edison's managers (if not Edison himself[*]) had promised the young inventor $50,000 (or the equivalent of $1.25 million today) for designing improvements to Edison's DC generators.[††] Tesla came up with twenty-four different improvements, mostly involving replacing the long magnets Edison used in his dynamos with shorter ones. These short-core magnets not only required fewer raw materials, they also improved the efficiency of the generators. Having successfully completed the task, Tesla is said to have confronted the manager about the $50,000 promise and to have been told that it was merely a joke (and that the Serbian simply didn't appreciate the sarcastic American sense of humor).[47]

According to the conventional account of most biographers, Tesla resigned in protest. But, there is evidence that the inventor continued to work for Edison, specifically helping to develop an arc-lighting system that would be better suited for outdoor illumination than Edison's dim incandescent bulbs. Though Edison had patented the basic plan for such a system as early as June of 1884, he charged Tesla with designing the details. However, by the time Tesla developed a working design, Edison had already entered into an agreement with the American Electric Manufacturing Company (AEM), whereby Edison's company would agree to install the AEM arc-lighting system for its outdoor lighting customers if AEM agreed to install Edison's incandescent system for its indoor lighting customers. The mutually beneficial deal meant Tesla's design was tabled. It was only after this affront that Tesla actually tendered his resignation.[48]

With Edison, Tesla was paid a salary of at least $10 a week (equivalent to about $1,000 a month today)—by no means a princely salary, but certainly not below-average for the time.[49] Still, upon leaving the Edison Company, Tesla found himself (again) dead broke and took to digging ditches for Western Union, which was running underground cables to link its main telegraph offices at 195 Broadway to the stock exchanges at Manhattan's southern tip.

**   Cheney references an earlier biography by John J. O'Neill in which Edison is quoted as telling Tesla that "there's fifty thousand dollars in it for you" if the Serbian could make the promised improvements to Edison's dynamos. [See O'Neill, *Prodigal Genius* (New York: David McCay Co., 1944), 93–95.] But Tesla refers only to "the manager" of Edison's New York offices and not to the man himself.

††   According to Cheney, Edison's people claim that it was Tesla who approached Edison with an offer to sell his patents for an AC generator for $50,000, which Edison declined. This story is unlikely since Tesla did not apply for a patent involving AC until 1888 and was not granted one on his "electro-magnetic motor" until at least three years after this incident is said to have occurred. See Cheney, *Tesla*, 57.

Illustration of an early arc-lighting system used for street illumination

Fortunately, the foreman who hired Tesla was a friend of Alfred S. Brown, a self-made electrician who had worked his way up the corporate ladder to become the head of Western Union's telegraph service for the entire New York Metropolitan District. The foreman introduced Tesla to Brown, who soon learned of the young inventor's efforts to develop a working AC generator. Sensing a lucrative opportunity—but knowing that Tesla lacked the business savvy to compete in the rough-and-tumble world of New York entrepreneurship—Brown solicited the help of New Jersey lawyer Charles Peck. Peck had a history of making a hefty profit by setting up competing businesses that Western Union would buy out in order to retain its virtual monopoly on the telegraph market. Peck was also good friends with William Stanley, a brilliant electrician whom the illustrious American entrepreneur George Westinghouse had hired to work on an AC power system the industrialist had bought from two European engineers (the French inventor Lucien Gaulard and the British engineer John Dixon Gibbs) after witnessing their successful long-distance transmission of AC power at the 1884 Turin International Electrical Exposition (see p. 103). Brown and Peck decided to underwrite Tesla's work, and, in the fall of 1886, established a laboratory for him on Liberty Street in lower Manhattan. Eventually, they incorporated their enterprise as the Tesla Electric Company.[50]

The War of the Currents

At the Liberty Street laboratory, Tesla worked feverishly on a new design for a motor using alternating current (AC). While the first electrical systems were powered by direct current (DC),

by the time Tesla was working on his motor, AC was in widespread use. Westinghouse had already established a company that was selling AC systems based on Gaulard and Gibbs's transformers. By the time Tesla had filed for a patent on his design, in fact, Westinghouse was claiming to have "sold more central station[s]…on the alternating current system than all of the other electrical companies in the country put together on the direct current system."[51] Still, the Gaulard-Gibbs design had some significant drawbacks. For one thing, it required stepping-up voltages to dangerously high levels. For another, to do so it depended on a transformer design then owned by Edison (see p. 128).‡‡ In fact, it was to get around the latter problem that Westinghouse hired Stanley. The wily entrepreneur tasked the electrician with modifying the Gaulard-Gibbs system just enough to be patentable and, thus, avoid a prolonged court battle with Edison.

For his part, Edison had invested quite a lot in developing and marketing an electrical system based on direct current (DC). Although it could not be transmitted very far, DC operated at lower voltages and was useful for powering applications, like freight elevators or cable cars, where the generator was located close to where the electricity was being used. Moreover, at the time Edison developed his system, there were no practical motors that used AC power. Most motors and other electrical implements evolved from systems designed to run on batteries, which only supply DC.

But Edison was a shrewd businessman, arguably more adroit at marketing than invention.

---

‡‡   Seifer speculated that Edison may have, in fact, purchased manufacturing options on the ZBD design for the sole purpose of blocking AC competitors. See Seifer, *Wizard*, 45.

He launched a publicity campaign to discourage the use of AC power. His most vociferous claim was that AC power required dangerously high voltages. In an 1886 letter to one of his plant managers, he speculated that "Westinghouse will kill a customer within six months after he puts in a system of any size."[52] Though Edison's concerns were clearly sincere, he expressed them through shameless displays of animal cruelty, going so far as to employ the electrician Harold Brown to tour the country demonstrating the dangers of AC power by electrocuting horses and dogs (and at least one elephant).[§§]

Westinghouse and Edison faced a third competitor, Elihu Thomson, who, together with his high school colleague, Edwin J. Houston, had formed the Thomson-Houston Electric Company.[¶¶] Thomson was born in Britain, but had immigrated to America in 1858. An accomplished electrical engineer, he had already patented several designs for electric arc-lighting lamps as well as his own AC power distribution system.[53] Like Westinghouse's system, Thomson's AC lighting system relied on very high voltages, a fact that worried the cautious electrician. In an 1887 lecture before the newly formed American Institute of Electrical Engineers (AIEE), Thomson sided with Edison, arguing that the voltages required to transmit AC power presented too great a public safety risk.[54] Thus, fully a year before Tesla had even patented his famous design for an AC motor, the War of the Currents was already in full swing.

Shortly after Tesla received a patent on his AC motor in 1888, he, Peck and Brown set about trying to sell rights to the design. Late in July, the inventor traveled to Pittsburgh to visit Westinghouse's factory and to negotiate terms for the sale not only of his AC patents, but options on up to forty patents Tesla had filed for various electrical components and lighting systems. When all was said and done, Westinghouse agreed to purchase the patent rights for an initial outlay of $75,000, two hundred shares of stock in Tesla's company, and $2.50 per watt in royalties fees. The total came to more than $250,000 over fifteen years, $180,000 of which was to come from royalties alone. In 1891, however, Tesla waived the royalty

---

§§ The term "electrocution" was coined by members of the press covering these gruesome spectacles. In service to Edison, Brown also built America's first electric chair, based on a design by Dr. Alfred Southwick, a dentist from Buffalo, New York.

American entrepreneur George Westinghouse

¶¶ In 1892, the Thomas-Houston Electric Company would merge with Edison's company to become General Electric (GE).

Topsy: The Shocking Execution of a Circus Elephant

One of the more gruesome events associated with the famous War of the Currents happened well after the war was won. A little after noon on January 4, 1903, a circus elephant named Topsy gained the unfortunate title of first elephant ever successfully executed by alternating current. In fact, there had been at least one previous attempt to electrocute an elephant, Jumbo II, at the 1901 Pan-American Exposition in Buffalo, but the twenty-two hundred volts employed appeared to do little more than unnerve the great beast. He eventually had to be poisoned with large doses of potassium cyanide.

The owners of Forepaugh Circus determined that Topsy—like Jumbo II before her—was too dangerous to continue working because the elephant was involved in a couple of violent episodes. In one incident, Topsy trampled to death a former circus employee, James Fielding Blount, who had jabbed the lit end of a cigar into the tip of her trunk after she refused to respond to his drunken torments. The New York newspapers that covered the event almost universally blamed Blount for having antagonized the elephant. In another incident, the police had to take Topsy into custody after witnessing one of her untrained handlers, Frederick "Whitey" Ault, stabbing the elephant in her trunk and between her eyes with a pitchfork when she failed, after several tries, to drag a multi-ton airship from one place to another. Ironically, when the police attempted to arrest Whitey for abusing Topsy, he cowered behind her for protection (which she dutifully offered). After much taunting by the perpetually inebriated handler, the police were finally able to cuff Whitey. But the damage had been done, especially after the *New York Times* ran the embarrassing headline, "Elephant Terrorizes Coney Island Police." James Bailey (of "Barnum & Bailey" fame), who owned the Forepaugh Circus at the time, decided that Topsy was not worth the bad publicity and would be put to death.

It is unclear how it was decided that Topsy would be electrocuted. But after the failed attempt on Jumbo II, Bailey turned directly to Edison—whose company supplied electricity to Coney Island—to oversee the event. Interestingly, when the War of the Currents was fully raging, Edison had sought (unsuccessfully) to demonstrate the danger of alternating current by electrocuting yet another Forepaugh elephant, Chief, in 1889. And there is also evidence that an agent of Edison's may have approached a commission determining the manner of execution of a violent elephant named Tip in 1894. Although Edison was not present at the Coney Island theme park where Topsy's execution was witnessed by a record crowd, he did send a film crew to document the morbid event for one of his early "kinetoscope" film reels; the aptly titled "Electrocution of an Elephant—Thomas Edison" can be viewed on YouTube and various other sites online.

Topsy's electrocution, documented by Thomas Edison

agreement in order to help keep Westinghouse from filing for bankruptcy (see p. 109). The inventor, therefore, never pocketed the vast majority of this potential windfall.

Niagara Falls and the Chicago World's Fair

On the Willamette River in Oregon, between Oregon City and West Lynn, lies the largest waterfall in the Pacific Northwest. Nearly thirty-one thousand cubic feet of water flow over Willamette Falls every second. In 1888, the Willamette Falls Electric Company (which would become Portland General Electric) was formed to find a way to harness the energy of the falls to generate electricity. Initially the company employed four turbine-driven dynamos of Edison's design to transmit DC power over a fourteen-mile-long power line to Portland. At the time, the Willamette-to-Portland line was the single longest DC transmission line in the United States. Not a year had passed, however, when the power station at Willamette was destroyed in a flood. Rather than rebuild the DC system, the company's investors chose to go with AC power and install Westinghouse's experimental generators. Their decision was to mark the beginning of the end of Edison's DC system.

In 1890, the Niagara Falls Power Company formed the International Niagara Commission, a panel of electricity experts and financiers led by Sir William Thomson (better known as "Lord Kelvin"). The Commission was tasked with analyzing various proposals for harnessing the energy of Niagara Falls to generate electricity. The impetus for the project was as much ecological as it was industrial. Early Niagara

settlers had diverted part of the upstream flow of the Niagara River to makeshift canals designed to power waterwheels. Before long, industrial mills of all sorts began popping up around the falls. By the 1870s, both sides of the river were dotted with waterwheels, all tapping the flow to power various industrial plants, and by the 1880s, much of the falls was obscured by concrete retaining walls punctuated with occasional openings for dirty mill discharge water. The blight spawned the Free Niagara Movement, one of the nation's first environmental efforts. Among its goals was the replacement of the waterwheels with hydroelectric turbines and the restoration of the river's natural aesthetics.[55]

In June 1890, the International Niagara Commission met in London and decided to offer a monetary prize for the best plan to transmit electricity from the falls to the burgeoning city of Buffalo, New York. In all, fifteen Europeans and four Americans submitted entries, including a proposal from Edison for the long-distance transmission of DC power, as well as a proposal by Westinghouse to use compressed air as a transmission system.*** Although the Commission bestowed awards on eight of the nineteen proposals, none was deemed worthy of practical application and the panelists continued to look for better transmission plans.

Prompted by the overwhelming success of the Paris World's Fair in 1889, Frankfurt decided to host its own International

---

*** Westinghouse did not propose Tesla's AC system—nor did he allow any of his engineers to submit such a proposal—because he didn't want to give away plans for the Tesla design for the paltry $20,000 in award money the Commission was offering.

Electro-Technical Exhibition in 1891. The premier feature of the exhibition was the successful long-distance transmission of electricity to Frankfurt's city center from a generating station some 108 miles away in Lauffen am Neckar. Although plans for such a system had been in the works since 1886, opinion was divided over what type of electricity should be used. Planners ultimately selected a configuration designed by Russian engineer Mikhail Dolivo-Dobrovolsky and British electrician Charles Eugene Lancelot Brown that used a three-phased AC power system eerily similar to Tesla's design. Dolivo-Dobrovolsky and Brown's results were impressive enough to settle the War of the Currents once and for all, at least in Europe.

In attendance were half-a-dozen members of the International Niagara Commission as well as Westinghouse and several U.S. entrepreneurs. Upon returning to America, the financial backers of the Niagara Falls Power Company sent Edward Dean Adams, president of the construction company chosen to build the Niagara

Willamette Falls hydropower system (1888)

power plant, to Europe to confer with Dolivo-Dobrovolsky and Brown directly.[56] For his part, when Westinghouse returned to the U.S., he invited a prominent member of the Commission, the British engineer George Forbes, to tour the Westinghouse plant in Pittsburgh and see Tesla's AC power system for himself. As a result, Forbes convinced the Niagara Falls Power Company to go with the Westinghouse design. Construction of the three-phase AC power system began in 1895. By November 1896, the first commercial long-distance transmission of electricity began as

power from the Niagara plant lighted up the city of Buffalo.

Besides Dolivo-Dobrovolsky and Brown's remarkable display and the commercial operation of the AC system at Niagara, any lingering doubts that the Westinghouse system was superior to Edison's DC-based system were completely laid to rest when Westinghouse won the contract to provide electrical illumination for the World's Columbian Exhibition in Chicago in 1893. Westinghouse's success was due in no small part to the fact that he took a

Niagara Falls blight (circa 1880s)

loss of several tens of thousands of dollars in order to underbid competing proposals from Edison. In retaliation, Edison refused to let Westinghouse light the fair using the Edison-designed incandescent bulbs. To avoid patent infringement, Westinghouse had just a few months to design and manufacture a slightly different bulb. Edison's petulance failed and Westinghouse's efforts paid off. By lighting the Chicago Exhibition with the polyphase power system designed in part by Tesla, Westinghouse effectively won the War of the Currents in America and the world turned its full attention to developing AC power.

Grand Wizardry and the Great Fire

While Westinghouse was shrewdly out-maneuvering Edison, Tesla was achieving nearly cult-like status through a series of over-the-top demonstrations. At the end of February 1893, he attended the National Electric Light Association convention in St. Louis. Lecturing before a standing room–only crowd of over four thousand, Tesla performed for the first time some of the displays of electrical engineering for which he would become famous. He illuminated fluorescent tubes wirelessly, wielding them like industrial-era light sabers before the stunned crowd. Charging his body with high-voltage AC, his skin became luminous and streams of lightning flew from his fingertips. Among those in attendance was Thomas Commerford (T. C.) Martin, the famed journalist and electrician who was covering Tesla's St. Louis lectures for *Electrical Engineer*.

When word reached beyond St. Louis, Tesla achieved stardom, and not only among the scientific community. Returning from the trip, he completed his citizenship requirements between press calls and social engagements. The *New York Herald*, quick to capitalize on the inventor's rising popularity, published a full-length feature article describing Tesla as "the greatest living electrician."[57] In the article, Tesla recounted his childhood and his "invention" of the rotating magnetic field. More articles would follow, including features in *McClure's*, *New Science Review*, *Outlook*, and the *New York Times*.

For perhaps the first time in his life, Tesla was reaping the monetary benefits of his efforts. He began dining regularly at Delmonico's in New York City, where the rich and powerful could see and be seen. He befriended Robert Underwood Johnson, the famed editor of the journal *Century*, and his wife Katherine. Through the Johnsons' social connections, Tesla met prominent personalities like Mark Twain, naturalist John Muir, writer Rudyard Kipling, and Theodore Roosevelt (then merely a New York City mayoral candidate).

Tesla was able to parlay his newfound celebrity into a new company with the financial backing of two of the Niagara project coordinators. Since his AC system was gaining widespread commercial success (though Tesla profited little from it), the inventor set his mind to improving electrical lighting systems and discovering how to use electrical oscillators to tap into the resonant frequencies of the earth and its atmosphere. Unfortunately, Tesla's innovations seemed constantly stalled in the experimental stages, rarely being perfected for practical use. One column in *Electrical World* even lamented that "one is naturally disappointed that

nothing practical has yet proceeded from the magnificent experimental investigations with which Tesla has dazzled the world."[58]

Despite the lack of commercial success, Tesla continued to file patent applications at a dizzying rate—at least fifty-seven different patents between 1886 and 1895, including designs for new incandescent and fluorescent lamps, mechanical and electrical oscillators, capacitors, and a resonant transformer that would come to be known as the Tesla Coil. This feverish tempo, however, began to take a physical toll on the inventor, whom one reporter for *Outlook* described as having "reached the limits of human endurance."[59] Indeed, although the Serbian towered well over six feet tall, his weight rarely exceeded a paltry 142 pounds.[60]

On March 13, 1895, during one of the rare times when Tesla actually slept, his laboratory on South Fifth Avenue burned to the ground. Although Tesla maintained a stiff upper-lip in public, privately he was devastated, even suffering a physical collapse. While his financial backers expressed their sincere regrets, many assumed the inventor was well insured and would be able to pick up at another location.[61] But Tesla, not one to be stalled by practical concerns, had never obtained insurance, and when Westinghouse's company began billing him for lost equipment, it quickly became apparent that the inventor would need to raise additional revenue to restore his laboratory.

Fortunately, Edward Dean Adams, president of the Cataract Construction Company, which had built the Niagara power plant, proposed to form a new company with Tesla. For $40,000 and an initial subscription of 20 percent of the company's stock offering, Adams would help Tesla get back on his feet. Tesla agreed, found a new location on East Houston Street, and quickly began stocking the space with the equipment necessary to continue his research.

Escape to Colorado

In 1854, the British scientist and Anglican priest William Whewell first theorized that there might be intelligent life on the planet Mars. By the late 1800s, interest in Martian life exploded as the advent of new telescopes enabled scientists to observe what appeared to be canals on the surface of the planet. In 1894, an American businessman and astronomer, Percival Lowell, made detailed descriptions of these canals in the journal *Nature*. Tesla should not have aroused such a stir, therefore, when, during an interview

for *Electrical World* in 1896, he suggested the possibility of "beckoning Martians" through the transmission of electrical waves.[62]

Wireless transmission of electrical waves was Tesla's new obsession. While he described his system of AC power as "a long-sought answer to pressing industrial questions" that came to him suddenly, "in a psychological moment," Tesla admitted his inventions for the wireless transmission of information—and, he claimed, of electricity—evolved from small steps intended "to improve the present designs."[63] Despite his alleged ability to conjure fully developed apparatuses in his mind, and to deduce new innovations from sheer force of reason, the inventor conceded that designs for his wireless system were "the product of labors extending through years."[64]

The fire in Tesla's Fifth Avenue laboratory had destroyed all of the equipment the inventor had been using to test the transmission and reception of electrical waves ever since he had first described the process in detail during a series of lectures in 1893.[65] It did not, however, appear to thwart his progress. During 1896 alone, he was granted eight different patents for electrical equipment associated with the wireless transmission of electromagnetic pulses, mostly oscillators used to generate high frequency waves. By the turn of the century, Tesla had applied for and received over thirty-three patents having to do with the wireless transmission of electromagnetism.[66]

Like most physicists, Tesla was aware of the tendency of mechanical systems to oscillate with greater amplitude when excited at certain frequencies. By this method of resonance, small forces applied to "shaking" the system could produce relatively large oscillations as the system stores and enhances vibrational energy. Tesla theorized that the earth was essentially one big oscillator, vibrating—like all things—at a certain rate. If he could find the resonant frequency of the earth, moreover, he surmised that he could use the earth's own electromagnetic currents to transmit information and energy throughout the globe.

Apparently undeterred by the earlier lab fire, Tesla used his new Houston Street lab to experiment with the concept of resonant frequencies, building and testing mechanical oscillators capable of bringing the whole edifice down upon him. During one famous experiment, Tesla attached an oscillator to the center support beam in the basement of the Houston Street building, and made adjustments in the frequency of vibrations until the beam began to emit a low ring. According to an assistant, while Tesla was attending to something else for a moment, the oscillation "attained such a crescendo of rhythm that it started to shake the building, then it began shaking the earth nearby," as well as other buildings. Machinery in the basement began to move across the floor and "the only thing which [sic] saved the building from utter collapse was the quick action of Dr. Tesla in seizing a hammer and destroying his machine."[67] (Later, Tesla would tell a reporter from the *Brooklyn Eagle* that he could put an end to mankind by splitting the earth using this same mechanism.)[68]

Tesla's Martian speculations and oscillation experiments raised some concern among his financial backers, who were more interested in profiting off the inventor's arc-lighting system than his grand plans for a "world telegraphy system." Having spent most of Edward Dean

New York World article on the fire at Tesla's laboratory (December 9, 1906)

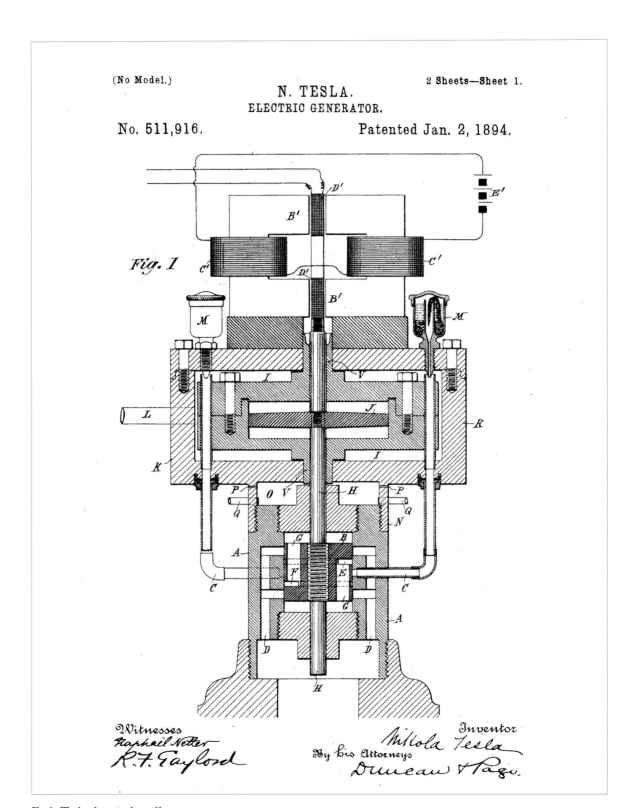

N. TESLA.
ELECTRIC GENERATOR.

No. 511,916.　　　　　　　　　Patented Jan. 2, 1894.

Fig. 1

Witnesses
Raphael Netter
R. F. Gaylord

Inventor
Nikola Tesla
By his Attorneys
Duncan & Page.

Early Tesla electrical oscillator

Adams's investment on these experiments, Tesla nevertheless leveraged Adams's support to solicit the financial backing of the wealthy attorney and businessman William Waldorf Astor. Astor was cautious. He warned the ambitious inventor that he was "taking too many leaps" and wanted to see Tesla demonstrate "some success in the marketplace before you go off saving the world with an invention of an entirely different order…"[69] Once Tesla assured him that he already had contracts for his lighting system pending with multiple European firms, Astor relented and on January 10, 1899, signed an agreement whereby he would invest $100,000 for five hundred shares of Tesla Electric Company stock and be named director of the company's board. This new agreement in effect diluted Adams's interest in the company.[70]

With Astor's money, Tesla moved into the businessman's luxury hotel, the Waldorf-Astoria—at the time the tallest hotel in the world and home to many of New York City's most prestigious residents. He also began plans to set up a new laboratory in Colorado Springs. Unbeknownst to Astor, Tesla had been quietly scouting the nation for an ideal spot for his new "Experimental Station."[71] He had visited Colorado Springs as early as 1896, to meet with his patent attorney at the time, Leonard Curtis (who had relocated for health reasons), and to conduct experiments in wireless transmission. Writing to Tesla from the area, Curtis suggested that the inventor could utilize Colorado's altitude and relative remoteness to avoid potential disturbances in wave propagation from the high buildings and general bustle of the city. In a letter of reply, Tesla indicated

his need to conduct tests using large amounts of electricity and his desire to keep those tests a secret. Curtis quietly approached the El Paso Power Company to arrange for Tesla to take any excess power the utility generated for free.[72] Once Astor left for Europe, Tesla laid the groundwork for his new laboratory and began shipping equipment to Colorado Springs.[73] After a short stop-off to make a presentation to the Chicago Commercial Club, the inventor arrived at the new Colorado laboratory on May 18, 1899. By the time Astor returned to the city nearly a month later, Tesla was gone, leaving his twenty-five-year-old assistant George Scherff to head the Houston Street laboratory.

Tesla attempted to keep his activities in Colorado Springs as confidential as possible.

William Waldorf Astor

Tesla's Experimental Station in Colorado Springs (1899–1900)

He employed a local carpenter, Joseph Dozier, to build a laboratory consisting of one large room, nearly thirty-six hundred square feet in size, with two small offices in the front. Though Dozier had installed a single large window, Tesla had it boarded up. Dozier was directed to build a wooden fence around the entire facility, on which Tesla hung signs reading: KEEP OUT, GREAT DANGER. Above the laboratory's entrance he penned a phrase from Dante's *Inferno*: "Abandon hope, all ye who enter here."[74]

We know now that Tesla was building what he called a "magnifying transmitter," which took the five hundred-volt electricity fed from the El Paso Electric Company, stepped it up to nearly forty thousand volts using a standard Westinghouse transformer, and then stepped that up to levels in the millions of volts using Tesla's own induction coil (see Chapter Five). Although the idea was to use the facility to perfect a system of wireless telegraphy, Tesla was also pursuing his ideas about resonant frequencies by sending high-voltage electrical signals through the ground and the atmosphere.

Tesla configured a relatively crude device (essentially a telephone receiver) designed to beep when it detected electrical disturbances over a thousand miles away. One rare evening, while Tesla was alone in the laboratory, the receiver signaled three steady beeps. The inventor was surprised to detect a signal with "such a clear suggestion of number and order."[75] At first, Tesla attributed it to periodic oscillations from a thunderstorm he had been monitoring for electromagnetic waves. Over the next several days, however, he became convinced that the signal may have originated from a neighboring planet and be evidence of an alien intelligence. In a holiday letter penned some weeks later to the local Red Cross Society, Tesla suggested that the electrical actions he had observed might portend a message from another world.[76] Recounting the event in a 1901 article for *Colliers* entitled "Talking with the Planets," Tesla announced the growing feeling that "I have been the first to hear the greeting of one planet to another."[77]

Within the popular press, Tesla's announcement was met with the kind of praise reserved for a god. Writing for the *Philadelphia North American*, the journalist Julian Hawthorne (son of famed author Nathaniel Hawthorne) wrote that Tesla was among the rare men, both scientist and poet, who may walk upon the ground, but whose head is among the stars: "Men of this mark are rare. Pythagoras was one; Newton must have had a touch of the inspiration; in our own times Tesla is the man . . ."[78]

His fellow scientists, on the other hand, were not so complimentary, suggesting that Tesla's fantastic claims were motivated more by self-promotion than by scientific discovery. One critic, known only as "Mr. X," writing in *Popular Science Monthly*, suggested that Tesla merely wanted to "figure in the newspapers." Admitting that everyone would be greatly interested in Tesla's signals if, indeed, they were transmitted from Mars, the anonymous writer concluded quite accurately that, "unfortunately, he has not adduced a scrap of evidence to prove it."[79]

Rather than the recipient of interplanetary messages, Tesla more likely had inadvertently intercepted one of the early radio signals Guglielmo Marconi was transmitting across the English Channel. In fact, on July 28, 1899—the very date on which historians suggest Tesla received the signals—Marconi was demonstrating his early radio apparatus to members of the British Admiralty and the French Navy. Moreover, his test signal was the letter "S" in Morse code, three steady dots that correspond exactly with the three beeps Tesla observed at his Colorado Springs laboratory some forty-seven hundred miles away.[80]

Losing Astor, Gaining Morgan

By January of 1900, Tesla had exhausted all
of the funds he had received from Astor and
returned to New York City, leaving unpaid
guards to watch over the Colorado laboratory.
He also left a large outstanding bill from the
El Paso Electric Company, incurred when one
of his experiments shorted out the local electric
grid and caused a total blackout in Colorado
Springs. Angered that Tesla had gone to
Colorado Springs without meeting his promise
to commercialize his fluorescent lighting system
and mechanical oscillators, Astor was not about
to advance Tesla any more money.[81] Tesla then
approached George Westinghouse for a loan,
suggesting that royalties he was to be paid by
companies in England could serve as collater-
al. Though he showed no interest in investing in
Tesla's wireless shenanigans (and worried that
Marconi had already beat him to the punch),
Westinghouse nevertheless advanced the inventor a few thousand dollars.[82]

At the turn of the century, industrial giant
J. P. Morgan was facing a bit of an existen-
tial crisis. His consolidation of the American
steel industry was nearly complete and he was
making unprecedented amounts of money.
By reducing labor costs and exploiting econo-
mies of scale, his company, U.S. Steel, would
become the first billion-dollar corporation the
world had ever seen.[83] Nevertheless, Morgan
was itching to expand his investments. On
the day after Thanksgiving of 1900, the steel
mogul and the indebted inventor met privately in Morgan's study. It seems Tesla's friend, the
esteemed architect Stanford White (who had
just designed the triumphal arch in New York

Guglielmo Marconi

City's Washington Square), had told Morgan of
Tesla's plans to build a wireless tower to transmit messages across the Atlantic.[84] Morgan
was looking for a way to signal steamships at
sea and obtain instantaneous quotes from the
New York Stock Exchange while traveling in
England. Though Tesla tried to sell the financier
on the vague promise of remote photography
(essentially today's television) and his "world
telegraphy system," Morgan was interested in
starting small, $100,000 for two towers, one
on each side of the Atlantic. Like Astor, however, he worried that Marconi's system would
render Tesla's obsolete. Tesla countered that
Marconi was engaged in "mere child's play...

J. P. Morgan, Sr.

the alternating current motor for which he had received a patent in 1888. One of Tesla's letters to Morgan, in fact, contained the cryptic suggestion that Tesla's patents on the AC system were on legally shaky ground compared to his patents on wireless transmission: "I beg you to bear in mind that my patents in this still virgin field, should you take hold of them . . . will command a position which, for a number of reasons will be legally stronger than that held by those of my own discovery in power transmission by alternating current."[86]

Tesla did not receive an immediate reply. His fiscal situation growing desperate, the inventor arranged to meet with Morgan around Christmas of 1900. Tesla again stressed the activity of wireless competitors and the relative strength of his own patents. Eventually, fatigued by Tesla's persistence, Morgan relented, offering the inventor $150,000 to erect a transmitter to send wireless signals across the Atlantic. In exchange, Tesla would give Morgan a 51 percent majority interest in the endeavor. But there was one catch. Morgan wanted to be a silent partner. He explained: "I want to be frank. I do not have a good impression of you. You abound in controversy, you are boastful, and aside from your deal with Westinghouse, you have yet to show a profit on any other creation."

Knowing of Tesla's previous arrangements, Morgan stressed that his $150,000 investment was firm and he would not be bilked to pay for the inventor's continuing research.[87] When the paperwork was written up, however, Morgan insisted on a 51 percent interest in Tesla's *patents* (in addition to his already agreed-upon stake in the wireless transmission tower),

using equipment designed by others" with no commercial viability whatsoever. By tapping the earth's natural frequencies, however, Tesla assured Morgan that his system "was the best for transmitting substantive information and insuring total privacy."[85]

Morgan was unconvinced, however, and told Tesla he would need time to consider any partnership arrangement. Tesla followed up with a letter on December 10th in which he laid out at length the flaws with Marconi's system of wireless transmission and the virtues of his own. Interestingly, it appears that even at this late date Tesla entertained doubts about his actual claim to invention of

including those associated with his lighting system. Of course, the latter agreement would impact Astor's interest in the Tesla Electric Company. So the inventor contacted Astor to solicit his involvement in the new scheme with Morgan. The industrialist was noncommittal and, notably, expressed concerns that Tesla did not have the fundamental patents on the wireless system he was proposing. Reading Astor's ambiguity in the best possible light, Tesla took the industrialist's silence as assent and signed the deal with Morgan in March of 1901. With Morgan's funding secured, Tesla promptly repaid Westinghouse for the $3,000 loan the inventor had lived off since returning from Colorado.[88]

Wardenclyffe and the World Telegraphy Center

While Tesla was working feverishly to seal a deal with Morgan and to find other investors, he embarked on a calculated campaign to publicize his plans for a World Telegraphy Center capable of transmitting messages around the globe at the speed of light. In mid-February 1901, he announced to reporters that he had perfected his system of wireless transmission and would be ready to send transatlantic messages before the end of the year.[89] To buttress his claims, he even leaked information about his agents scouting locations for a transmitting station in New Jersey and a receiving station in Portugal.

Anticipating that Tesla's endeavors would bring upwards of two thousand employees (all of whom would need housing), James Warden, a Long Island lawyer and real estate speculator, offered the inventor two hundred acres of a sixteen hundred acre parcel the businessman had purchased at the end of the Long Island Railroad's Port Jefferson line. Expecting to attract wealthy New Yorkers escaping the city's brutal summers, Warden christened the property Wardenclyffe. Tesla signed the deal with Morgan, accepted Warden's offer, and broke ground on his transmission tower by September, 1901.[90]

Tesla tapped his friend Stanford White to draw up the designs for the wireless station at Wardenclyffe. It was to include a single brick building, encasing over nine thousand square feet of working space, and a large tower situated 350 feet away. Below the tower, Tesla wanted to construct a well (with a spiral staircase) sunk almost ten stories underground. He also had White draw up plans for a model city stretching over eighteen hundred acres,

Stanford White

including stores, civic buildings, and housing for over twenty-five hundred workers.[91]

Despite the $100,000 figure he originally cited to Morgan as the cost to construct the tower, Tesla now estimated that the necessary funds would exceed $450,000.[92]

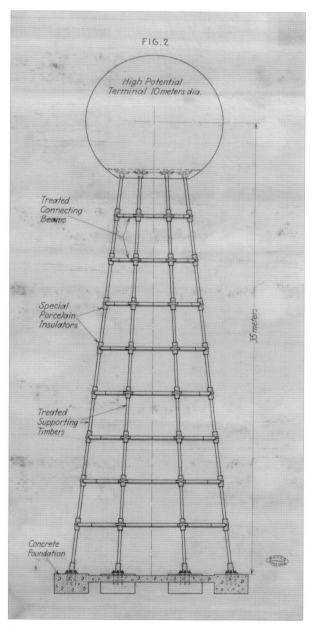

Wardenclyffe design

His calculations—based on his experiments at Colorado Springs—indicated that a tower sufficient to transmit his signals worldwide would need to be six hundred feet high, nearly two-thirds the height of the Eiffel Tower and taller than any building then-existing in America. The half-sphere terminal at the top would be sixty-eight feet in diameter and weigh fifty-five tons. Despite being warned that Morgan's $150,000 investment was firm, by mid-September, Tesla was begging the financier for additional funds. Morgan refused. Tesla would have to settle for a tower of only 187 feet, still tall enough to be seen with the naked eye from the other side of Long Island Sound, nearly fifty miles away.

Tesla's grand construction project did not go unnoticed. Marconi, who had scouted locations for his own transmitting tower on Long Island in 1901, hastened to transmit a transatlantic signal using his own wireless system. To make the feat more achievable, Marconi decided to construct a transmitting tower in Cornwall, England, and attempted to receive the signal via an aerial suspended from a kite in St. Johns, Newfoundland, the closest spot to England in North America. On December 12, 1901, his aerial picked up the familiar "beep-beep-beep" Morse code for the letter "S."[†††] Within days, the announcement of his achievement made the world's major papers, including the front page of the *New York Times* (above the fold!).[93]

The system appeared to work and Marconi was signing investors left and right; meanwhile Tesla was assuring Morgan that Marconi's

---

††† Interestingly, there was no physical evidence that the signals actually were received, and the only two people to witness the event were Marconi and his assistant, George Kemp.

system would never achieve commercial success. In an attempt to salvage the situation, Tesla met with Morgan and presented a plan to create a new company with the intention of expanding Wardenclyffe to its full potential, capitalized with $10 million raised from potential investors. The inventor even proposed to repay Morgan's initial $150,000 investment by selling bonds in the new company.[94] Morgan agreed to bring on new investors, but was not about to sink any more of his own money into the enterprise, especially after Tesla revealed that he would have to haul truckloads of coal out to Long Island to generate the massive amount of energy required to power the full version of his global transmission system.

In the meantime, unpaid bills stacked up. Tesla still had debts incurred in Colorado Springs. He owed Westinghouse nearly $30,000 for equipment to furnish Wardenclyffe and he was being sued by James Warden for failure to pay the property taxes on the two hundred acres he had donated to the inventor. Tesla sold off about $30,000 of his personal property and took out a loan from a Port Jefferson bank for another $10,000.

Bankruptcy and Descent into Madness (or Genius?)

By July 1903, the inventor made a desperate appeal to Morgan, admitting that he "was in a dreadful fix" financially.[95] By that time, however, the financier not only was concerned that he had not seen a penny of return on his investment after almost two-and-a-half years, but also that his only partner in the endeavor was losing his grip on reality. Moreover, even if he were eventually successful, Tesla's scheme appeared designed to provide the world with free energy, a plan that would likely bankrupt Morgan if it ever succeeded. According to John O'Neill, Tesla's first biographer, the rising Wall Street tycoon Bernard Baruch warned Morgan about this possibility: "Look, this guy is going crazy . . . he wants to give free electrical power to everybody and we can't put meters on that. We'll go broke supporting this guy."[96] Morgan apparently agreed.

For help, Tesla turned to his financier's mortal enemies, everyone from Charles A. Coffin, then-president of Westinghouse's rival General Electric, to Jacob Schiff, the investor who had tried (unsuccessfully) to wrest financial control of the Northern Pacific Railroad from Morgan in 1901.[97] But the desperation of the inventor's entreaties belied his persistent claim that his world-shaking innovations were on the brink of being realized. Moreover, the deal he had penned with Morgan provided little incentive for new investors. After all, Morgan, who owned 51 percent of the patent rights on any wireless transmission system, controlled whether those patents could be exercised and would pocket a disproportionate share of any profits.[98]

Tesla devised even wilder schemes to raise money. In 1902, he tried to form a new company—Tesla Electric and Manufacturing Company—to produce smaller forms of his inventions for use in scientific laboratories. But potential investors were put off by the high start-up costs. He attempted to commercialize a smaller version of his induction coil system as a room sanitizer (the ozone produced as a by-product of the coil's high voltages killed germs in the air). He even sought to capitalize on his fame in his home country, pleading unsuccessfully with a Serbian bank for a short-term loan.[99] Finally, the esteemed inventor was forced to peddle his intellect as an electrical engineering consultant and asked his friends Robert and Katherine Johnson for a list of "prominent and influential" people to whom he could send a prospectus.[100]

In desperation, Tesla continued his appeals to Morgan for funding to complete the Wardenclyffe facility. In a series of letters to the industrialist between 1903 and 1905, Tesla touted his unparalleled genius, claiming that he had "more creations named after me than any man that has gone before not excepting Archimedes and Galileo."[101] He also claimed, without empirical evidence, not only to have perfected a system by which electricity could be transmitted wirelessly across thousands of miles, but also that he alone possessed the genius to figure it out:

I have perfected the greatest invention of all time—the transmission of electrical

energy without wires to any distance . . . the long sought stone of the philosophers . . . I am the only man on the earth to-day who has the peculiar knowledge and ability to achieve this wonder and another may not come in a hundred years.[102]

Morgan, nevertheless, was unimpressed. The tycoon was facing congressional efforts to break up his steel monopoly and had little time to reply to Tesla's repeated entreaties.

Physically weary and psychologically defeated, Tesla suffered another nervous breakdown followed by a bout of cholera in the fall of 1905. His more eccentric quirks and foibles became even more pronounced. His handwriting became nearly illegible and the subject of his writing (when it could be read at all) more unhinged.[103] His inventing fell off as well: he failed to file a single patent application from late 1905 until well into 1909.

Ironically, it was during this period of turmoil that Tesla devised some of his most

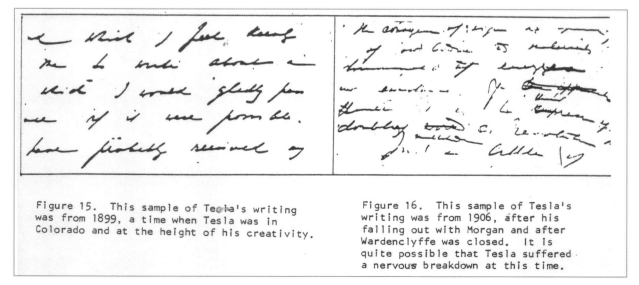

Figure 15. This sample of Tesla's writing was from 1899, a time when Tesla was in Colorado and at the height of his creativity.

Figure 16. This sample of Tesla's writing was from 1906, after his falling out with Morgan and after Wardenclyffe was closed. It is quite possible that Tesla suffered a nervous breakdown at this time.

Comparison of Tesla's handwriting before and after 1905

Patent sketch of Tesla turbine

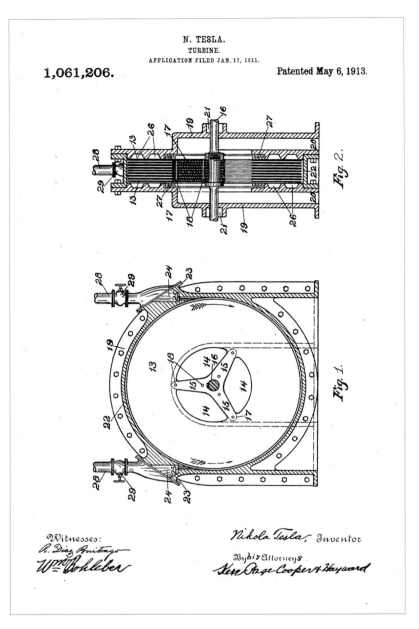

N. TESLA.
TURBINE.
APPLICATION FILED JAN. 17, 1911.

1,061,206.

Patented May 6, 1913.

Fig. 2.

Fig. 1.

Witnesses:
R. Diaz Buitrago
Wm Rohleber

Nikola Tesla, Inventor
By his Attorneys
Kerr Page Cooper & Hayward

fascinating (and least heralded) inventions. In 1909, when his patent drought finally ended, he filed claims on a bladeless turbine and pump that utilized the tendency of fluids in motion to adhere to surfaces (known as the "boundary layer effect") in order to spin a series of discs attached to a rotor. The "Tesla Turbine" was not successful commercially at the time primarily because it competed with bladed turbine designs already in widespread use. Moreover, the very efficiency of its operation caused the discs to spin at such high revolutions that it literally warped the metal. Essentially, Tesla's turbine worked too well for the materials then available.

Nevertheless, Tesla was confident he could make the turbine work and partnered with Joseph Hadley, an Alabama coal baron, to create the Tesla Propulsion Company and proceeded to rent office space in the Metropolitan Life Tower, the tallest building in the world in 1909.[104] Despite his past dealings, he also approached Astor for funding. The multi-millionaire declined and, several years later, drowned—along with more than fifteen hundred others—when the Titanic sunk on its maiden voyage in 1912. Tesla intended to solicit Morgan once again as well. But the financier was marred in congressional inquests until his sudden death in 1913. Two months after the funeral, however, Tesla approached Morgan's son and chief heir, J. P. Morgan, Jr. ("Jack"). Intrigued by Tesla's innovative design, but unwilling to risk much on an inventor with Tesla's consistently

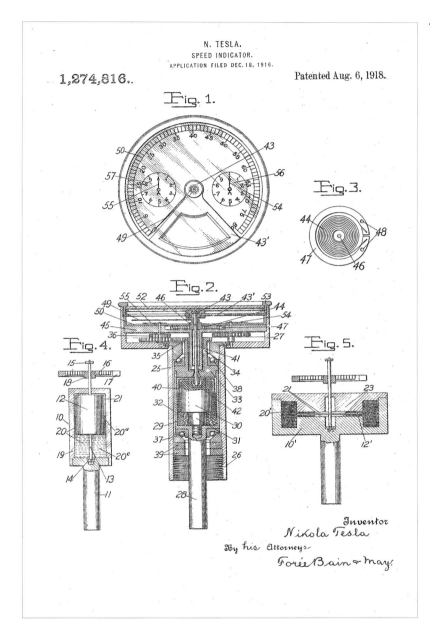

N. TESLA.
SPEED INDICATOR.
APPLICATION FILED DEC. 18, 1916.

1,274,816.

Patented Aug. 6, 1918.

Fig. 1.

Fig. 3.

Fig. 2.

Fig. 4.

Fig. 5.

Inventor
Nikola Tesla
By his attorneys
Forée Bain & May

had to rely on consulting services to keep his most recent company afloat. He moved his offices from the prestigious Woolworth Building to a cheaper facility just off Bryant Park, near the newly-erected New York Public Library building. When the city sued him for back taxes, however, the full extent of Tesla's poor financial situation finally became public. Despite his many patents and the commercial success of his AC power system, he was making only $300 to $400 a month (about $7,500 to $8,500 in 2014 dollars), barely enough to cover the company's expenses, and not nearly enough to pay off the dozens of creditors with judgments against him. Tesla admitted that he was living off credit; even the bill for his room at the Waldorf-Astoria hadn't been paid in years.[106] The court declared the inventor bankrupt and appointed a receiver to manage his finances.

disappointing record, Jack loaned the inventor $20,000. Tesla promptly used the money to move his new company to the even more impressive Woolworth Building (which, upon its completion in April of 1913, eclipsed the height of the Metropolitan Life Tower).[105]

Despite these last-minute injections of capital, by 1916, Tesla was once again broke and

Some historians have suggested that Tesla never really recovered from the mental breakdown he experienced following Wardenclyffe's failure and his subsequent financial troubles.[107] The inventor became somewhat of a recluse, spending the majority of his time exploring innovations in fluid dynamics. Unlike his earlier grand electrical schemes, Tesla filed patents for minor improvements in the construction of fountain and aquarium displays, various speedometers utilizing the adhesion and viscosity of gases, and even a novel design for a protective lightning rod.

The few times he did emerge, Tesla lived up to the reputation of mad scientist that started to plague him after he declared he had received messages from Mars. In 1917, he was awarded the Edison Medal by the American Institute of Electrical Engineers at a banquet held at the Engineers' Club off Bryant Park. As the official ceremony started, however, Tesla was nowhere to be seen and a frantic search ensued. Eventually, a senior engineer from Westinghouse found the erratic inventor in the park, earnestly feeding the pigeons.[108]

World War, Death Rays, and Pigeon Love

In mid-summer of 1914, a series of international alliances designed to deter the kinds of imperial aggressions that had plagued Europe for more than a thousand years suddenly plunged the globe into world war after Austrian Archduke Franz Ferdinand was assassinated by a Yugoslavian nationalist. Within weeks, all of the major European powers had coalesced along two factions and declared war on each other. Upon its declaration of war against Germany, Britain promptly cut all of the undersea telegraph wires that linked the two countries. Since the U.S. telegraph line depended on the undersea links between Britain and Germany, the only long-distance communication link left between Germany and the U.S. relied

exclusively on the few wireless radio stations erected along the Long Island and New Jersey coasts, all of which happened to be owned by German-controlled companies.

In an effort to challenge British dominance in radio, Kaiser Wilhelm II had, some years before the war, overseen the consolidation of German radio interests into one company, with the typically lengthy German name "Gesellschaft für drahtlose Telegraphie System Telefunken" (or "Telefunken," for short).[109] Recognizing the threat that German control of U.S. radio links might entail, the British convinced Marconi's American-based subsidiaries to sue Telefunken for patent infringement. Telefunken also understood the importance of the outcome of this litigation and soon employed the best U.S. physicists, including Tesla, to serve as expert witnesses. Thus, for $1,000 a month, Tesla was employed by Telefunken to defend a legal position he espoused anyway: that Marconi's radio patents were fraudulent. It was only after Tesla knew that German resources were being brought to bear against Marconi, however, that the inventor decided to file his own patent infringement suit against the Italian.[110]

By April of 1917, however, the legal case against Marconi became moot when the U.S. officially entered the war and the U.S. Navy hastily seized all radio stations (including those owned by the British).[111] To ensure that the U.S. military could pursue technological innovations unhampered by legal formalities, President Woodrow Wilson suspended all patent litigation for the duration of the war. Tesla's claim would have to wait. Tragically, by the end of the war, when Congress approved government payments for the use of any invention covered by U.S. patents, all of Tesla's radio patents had already expired.[112]

As if that weren't enough, Tesla had surreptitiously been licensing his radio patents to the U.S. Navy since 1912.[113] In preparation for the war, Tesla claims the Navy installed as much as $10 million worth of equipment based on his designs. With patent litigation suspended, however, there was little Tesla could do to force the Navy to pay up. Even worse, because the equipment apparently was being used to test a crude guided missile system, much of the work was classified and immune to patent litigation anyway.[114] Once again, Tesla's fortune was thwarted by extraordinary circumstances and poor timing.

Sometime in 1917, after maids at the Waldorf-Astoria repeatedly complained of foul odors and a shocking amount of pigeon feces, Tesla was evicted from the hotel room he called home for more than fifteen years. As he left, he was handed a bill for over $19,000 (equal to nearly half-a-million dollars today) in unpaid rent and room service.[‡‡‡]

The impoverished inventor turned to consulting in a desperate attempt to keep ahead of his mounting debts. Packing his papers and a few belongings, he set out for Chicago to fulfill a gig with Pyle National, a company

that, to this day, is one of the world's leading distributors of electrical switches. Before leaving, however, he sent a letter to Jack Morgan (to whom he still owed more than $25,000), ensuring the industrialist that he intended to repay his debt with the profit he would make on an invention that would "afford an effective means for meeting the menace of the submarine," leading many to speculate that Tesla had perfected long-distance radar.[115]

From 1917 until 1926, Tesla moved from one consulting gig to another, bouncing from Chicago to Milwaukee to Boston to Philadelphia.[116] The windfall Tesla had promised Morgan never came, however. During the course of the war, Washington had enlisted Thomas Edison's help in developing new military technologies, tapping him to lead the Navy Consulting Board, a kind of high-tech think tank that Edison (the consummate profiteer) had himself suggested to the Secretary of the Navy, Josephus Daniels, as early as 1915. Throughout the war, Tesla would float ideas in the popular press for newfangled military weapons, including cryptic references to a kind of long-distance death ray reported in one *New York Times* article (that, incidentally, incorrectly credited the inventor with winning the 1915 Nobel Prize in physics).[117]

Dreams, Death, and Final Accolades

At the close of World War I, Tesla was approached by Vladimir Lenin, leader of the newly-formed Soviet Union, apparently with the intention of employing Tesla's wireless transmission system to bring electricity to all of Russia (a goal Lenin felt was essential to the success of communism). Although

---

‡‡‡ Tesla had held off eviction from the Waldorf-Astoria by mortgaging the property at Wardenclyffe to the hotel's owner, George C. Boldt, in 1904. What was left of the actual tower was torn down and the scrap metal sold to pay off some of Tesla's creditors. By 1921, the courts awarded what remained of Wardenclyffe to the Waldorf-Astoria to help defray the costs of housing and feeding Tesla for nearly 16 years. Boldt eventually sold the facility to Brooklynite Walter L. Johnson, who subsequently leased it to a photo products company. After that company closed its doors, the New York Department of Environmental Conservation spent several years cleaning up accumulated silver and cadmium waste. In 2009, the State of New York put the property up for sale for $1.65 million. It was sold for an undisclosed price in 2013 to a nonprofit organization that intends to build a museum to Tesla on the site.

Illustration of Tesla's proposed particle beam weapon

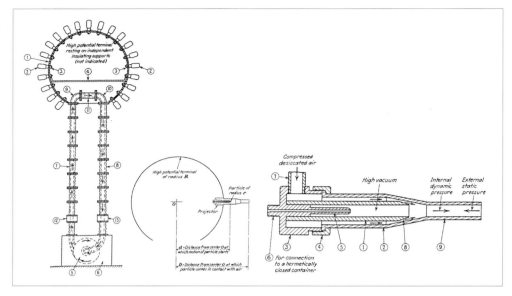

the revolutionary had little money, he promised to pay Tesla with "carloads of gold" that had been nationalized during the revolution. Tesla declined, however, saying that he first wanted to see what America would do with his invention.[118] Despite mounting debts and his personal bankruptcy, Tesla remained convinced that capitalist enterprise within the U.S. would eventually bring his innovations—including his dream of a world telegraphy system—to fruition.

Notwithstanding his reclusiveness after the war, Tesla continued to give interviews in the trade press in which he would present novel ideas and futuristic visions, many of which were eerily prescient. In a 1917 article for *Electrical Experimenter*, for example, he described a system of detecting ships at sea using powerful waves of electromagnetic energy bounced off the ships and reflected onto fluorescent screens.[119] Many speculate that he, therefore, anticipated—if not invented—radar using microwave transmission (see Chapter Six).[120]

In 1934, Tesla gave an interview to the *New York Times* which convinced an intrepid reporter to claim that the inventor had developed and tested an "apparatus which will send concentrated beams of particles through the free air, of such tremendous energy that they will bring down a fleet of 10,000 enemy airplanes at a distance of 250 miles…"[121] Although there is no concrete evidence that Tesla ever built such a weapon, in 1984, Tesla enthusiasts began circulating a paper (eventually confirmed as Tesla's own writing) in which the inventor outlined a method of projecting microscopic tungsten particles to nearly light speed. Noting that the weapon used accelerated particles instead of rays, Tesla wrote that, "this invention of mine does not contemplate any so-called 'death rays.'"[122]

Typically, Tesla never pursued these ideas commercially and even his attempts to reclaim profits from infringements on his past inventions brought nothing but financial ruin. In 1925, his own patent lawyer, Ralph J.

THE ELECTRICAL EXPERIMENTER

H. GERNSBACK EDITOR
H. W. SECOR ASSOCIATE EDITOR

| Vol. V. Whole No. 52 | August, 1917 | Number 4 |

Tesla's Views on Electricity and the War

By H. WINFIELD SECOR
Exclusive Interview to THE ELECTRICAL EXPERIMENTER

NIKOLA TESLA, one of the greatest of living electrical engineers and recipient of the seventh "Edison" medal, has evolved several unique and far-reaching ideas which if developed and practically applied should help to partially, if not totally, solve

interview and some of his ideas on electricity's possible rôle in helping to end the great world-war are herein given:

The all-absorbing topic of daily conversation at the present time is of course the "U-boat." Therefore, I made that subject my opening shot.

pacity of chief electrician for an electric plant situated on the river Seine, in France, I had occasion to require for certain testing purposes an extremely sensitive galvanometer. In those days the quartz fiber was an unknown quantity—and I, by becoming specially adept, managed to pro-

Nikola Tesla, the Famous Electric Inventor, Has Proposed Three Different Electrical Schemes for Locating Submerged Submarines. The Reflected Electric Ray Method Is Illustrated Above; the High-Frequency Invisible Electric Ray, When Reflected by a Submarine Hull, Causes Phosphorescent Screens on Another or Even the Same Ship to Glow, Giving Warning That the U-boats Are Near.

the much discust submarine menace and to provide a means whereby the enemy's powder and shell magazines may be exploded at a distance of several miles.

There have been numerous stories bruited about by more or less irresponsible self-styled experts that certain American inventors, including Dr. Tesla, had invented among other things an *electric ray* to destroy or detect a submarine under water at a considerable distance. Mr. Tesla very courteously granted the writer an

"Well," said Dr. Tesla, "I have several distinct ideas regarding the subjugation of the submarine. But lest we forget, let us not underestimate the efficiency of the means available for carrying on submarine warfare. We may use microphones to detect the submarine, but on the other hand the submarine commander may employ microphones to locate a ship and even torpedo it by the range thus found, without ever showing his periscope above water.

"Many years ago while serving in the ca-

duce an extremely fine cocoon fiber for the galvanometer suspension. Further, the galvanometer proved very sensitive for the location in which it was to be used; so a special cement base was sunk in the ground and by using a lead sub-base suspended on springs all mechanical shock and vibration effects were finally gotten rid of.

"As a matter of actual personal experience," said Dr. Tesla, "it became a fact that the small iron-hull steam mail-packets (ships) plying up and down the river Seine

The Electrical Experimenter article in which Tesla proposes a crude radar system

Hawkins, had to take the inventor to court in an attempt to collect over $900 in legal fees Tesla owed him. Rather than allow the impoverished (and very likely, soon-to-be-homeless) Tesla—now nearly seventy years old—to face public embarrassment, the Westinghouse Company quietly agreed to hire the inventor as a "consulting engineer" and to provide him with a monthly pension of $125.[123] Tesla used this paltry sum to rent a suite at the Hotel New Yorker, where he resided until his death.[124] (Apparently, however, he reneged on paying Manhattan Storage $297 for holding onto his belongings for several years, and when the company threatened to auction off Tesla's personal effects, his nephew— the Yugoslavian ambassador to the U.S.—was forced to pay off the debt.)[125]

ABOVE: Tesla in 1933, looking gaunt

LEFT: Tesla's favorite pigeon, "white dove"

Despite his ardent physical regimen, which included eating only boiled vegetables and keeping visitors at a distance of at least three feet, Tesla's health deteriorated quickly after his seventy-fifth birthday. Sometime in 1937, he was struck by a taxicab, but refused medical treatment. By 1942, he was mostly bedridden and even his mind was fading. There is some evidence of Tesla's creeping senility, including attempts to send messages to colleagues long-since dead.[126]

Tesla described one night when he was lying in bed and his favorite pigeon, a female bird named "white dove" (that the inventor claimed could find him anywhere in the city), flew into the window of his room at the Hotel New Yorker.[127] According to Tesla, "powerful beams of light" came from her eyes and then slowly diminished. Tesla believed that her death signaled the end of his life's work.[128] The inventor died in his sleep of an apparent heart attack a few months later.[129] He was eighty-seven years old.

Tesla's coffin being carried by pallbearers

More than two thousand people attended Tesla's funeral, which was held at the Cathedral of St. John the Divine on New York's Upper West Side. Eleanor Roosevelt sent her condolences, along with those of President Franklin Roosevelt, who on the very night of Tesla's death had delivered his State of the Union address predicting the Allied victory in World War II. Mayor Fiorello La Guardia eulogized the inventor in a radio broadcast (still available online at the Tesla Memorial Society of New York) in which he noted that Tesla, "a natural genius" and "one of the most useful and successful men who ever lived," nevertheless died in poverty.[130] An editorial in the *New York Sun* proclaimed that "his guesses were right so often that he would be frightening" and predicted that the world would probably appreciate the inventor's "superb intelligence" "a few million years from now."[131] The near deification of Tesla in the past several years, however, may be proof that the *Sun* was off, perhaps, by only a few million years. Tesla's life may be over, but his legacy remains unresolved.

"I think we have systematically and critically harmed ourselves and many young people by advising them not to try things. 'Be careful of being a jack of all trades and a master of none.' If you are a human being, you can attempt to do what other human beings have done. We don't understand talent any more than we understand electricity."

—MAYA ANGELOU, *ARIZONA REPUBLIC*, JANUARY 12, 2015

chapter three

Understanding Electricity

Few people spend time thinking about how electricity gets from the power plant to outlets in their homes and offices. And many, no doubt, prefer it that way. Most people are not very excited by the complexities of resistance, conductance, voltage, and current, and those that are, thankfully, pursue degrees in electrical engineering so the rest of us don't have to. While it is possible to appreciate the enormity of the discoveries for which Tesla is idolized without mastering the complexities of electrical engineering, it is helpful to have some basic knowledge about electricity and how it is transmitted in order to understand Tesla's true contribution to the history of electrical innovation.

What Is Electricity?

Albert Einstein once famously declared that "everything in life is vibration." This is one way to think about electricity (and it is probably how

OPPOSITE: A rare photo of Tesla demonstrating some of the electrical tricks he performed during public lectures

Tesla thought of it).* At its most basic, electricity is the movement of an electrical charge along a conductor (say, a wire). It is, in essence, a kind of moving vibration, a wave. It is easiest for most people to think of an electrical charge as individual electrons moving through a wire like water through a pipe. But this conception is not entirely correct. Just like a tidal wave is a vibration that moves a tremendous amount of energy through the ocean without moving a tremendous amount of actual water, electrons themselves do not have to move from one place to another in order to create an electrical current. Rather, what we observe as the flow of electrical charge is an impulse of energy that moves *through* a collection of electrons (and other charged particles) like a kind of vibration, passing from electron to

---

* In one famous quote attributed to Tesla, the inventor claimed, "if you want to find the secrets to the universe, think in terms of energy, frequency and vibration." Moreover, throughout his life he rejected the idea of quantum mechanics, believing instead that energy traveled by conduction—a sort of charged wave propagating through an unseen "ether."

electron.† Nevertheless, most people fall into the habit of talking about electrical current as a flow of individual electrons from one place to another (and this author is no exception).

To truly understand electricity requires thinking about the fundamental nature of matter. Matter is simply anything having mass, from planets composed of trillions and trillions of atoms to the subatomic particles that make up atoms themselves. One factor common to all matter, from the smallest subatomic particle to the largest galaxy in the cosmos, is that it has physical properties, measurable characteristics that describe its existence in various states. Density, for example, is a physical property of matter. Color is another physical property. Electrical charge is yet another.

The electrical charge of matter describes how it reacts when it is exposed to an electromagnetic field. For our purposes, there are only two possible manifestations of charge: positive or negative. We know from observation that matter with a positive charge will repel other positively charged matter, but attract negatively charged matter. Conversely, matter with a negative charge will repel other negatively charged matter, but attract positively charged matter. Opposites attract—simple enough.

Things start to get a little more complicated when exploring *how* matter can be charged and how that charge travels through—or over, or around—matter. To understand why many of the innovations often attributed to Tesla were so revolutionary, we must first understand the difference between electrochemical

energy and electromagnetic energy, as well as the difference between direct current and alternating current.

Discovering Electricity: Electrostatic and Electrochemical Energy

Long before George Westinghouse and Thomas Edison waged their famous "War of the Currents," people were well aware that touching certain fish would create a shock. The Greco-Romans were even known to use the shock of an electric fish as a kind of local anesthetic. Ancient peoples along the Mediterranean coast knew that this sensation was related to the shock that could be produced by touching a rod of amber that had been rubbed with cat's fur. The first systematic observations of the phenomenon, in fact, were made around 600 B.C. by Thales of Miletus, one of seven famous Greek sages. Because the rubbed amber had a tendency to attract human hair, chicken feathers and other like objects, much like magnetic rocks (known as "lodestones" at the time), Thales incorrectly believed that rubbing fur against amber caused it to become magnetic.

What Thales was observing was not magnetism, but the imbalance of electrical charge on the surface of a material—what we know as static electricity. Most materials are electrically neutral because they are made up of atoms that contain an equal number of positively charged protons and negatively charged electrons. However, some materials hold on tightly to their electrons, while others do not. In fact, some materials barely hold on to their electrons at all. When these materials rub against tight-holding materials, the weak-holding materials simply give up their electrons

---

† While most people refer to electricity as the movement of electrons, charge can also be carried by other particles such as protons and ions (a molecule with an imbalance between the number of protons and electrons).

altogether. An imbalance in electrical charge is created as one material loses electrons (and becomes positively charged) while the other material gains electrons (and becomes negatively charged). When the two materials are separated, they retain this imbalance.

Because nature seeks balance, charged materials behave in some interesting ways. Say a material with an excess of electrons (negatively charged) comes near a material with a deficit of electrons (positively charged). Nature will compel the materials to share electrons. The result is an attractive force some enterprising marketers of fabric softener have come to refer to as "static cling" but what is more accurately described as an electrostatic charge. When

LEFT: Thales of Miletus

BELOW: Historical illustration of Franklin's kite experiment

electrons jump from one material to another—a phenomenon known as "static discharge"—it usually creates a spark or shock. Benjamin Franklin deduced that lightning was essentially static discharge on a much larger scale when he famously attached a metal key to the dampened string of a kite flown (dangerously) into an approaching thunderstorm.‡ He was right.

Static discharge occurs not only on the surface of materials, but also *within* materials,

in an electrochemical reaction known as reduction-oxidation (or "redox," for short). The chemical process of a substance gaining an electron, strangely enough, is known as "reduction," while the chemical process of a substance losing an electron is called "oxidation" (even though the reaction might not involve oxygen at all). This might seem counterintuitive at first. But, to help keep it all relatively straight, students of electricity have devised the simple mnemonic *OIL RIG* (*O*xidation *I*s *L*oss; *R*eduction *I*s *G*ain). Oxidation and reduction always occur in pairs; as one substance loses an electron (oxidizes), the other gains an electron (reduces)—hence the name "reduction-oxidation." Just as certain materials can become charged with static electricity

---

‡   Contrary to popular lore, Franklin's son William flew the kite (if it flew at all), while Franklin retreated to a barn to observe a Leyden jar (an early contraption used to harness an electrical charge) attached to the end of the kite string. It is also unlikely that the kite was struck by lightning as the resulting shock would undoubtedly have killed or injured his son. However, even without being struck, the string and attached key gathered an electrical charge, which may have accounted for the minor shock the elder Franklin received as he passed his hand close to the electrified key.

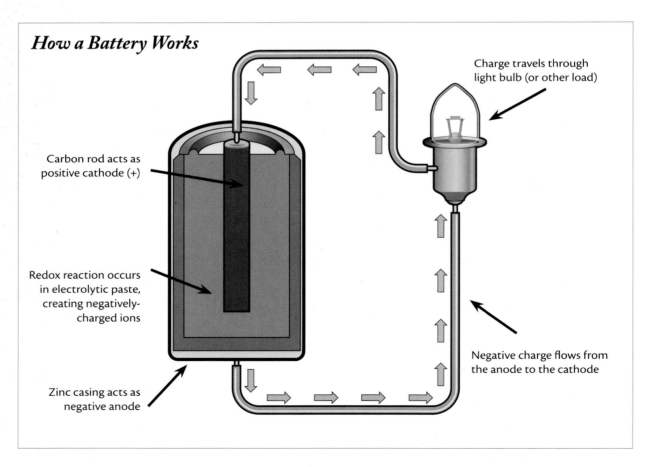

How a Battery Works

Charge travels through light bulb (or other load)

Carbon rod acts as positive cathode (+)

Redox reaction occurs in electrolytic paste, creating negatively-charged ions

Negative charge flows from the anode to the cathode

Zinc casing acts as negative anode

Ancient Electrochemistry: The Baghdad Battery

In 1938, as the Nazis planned their attack on Czechoslovakia, German archaeologist Wilhelm König was appointed director of the Baghdad Museum in Iraq. While digging around the museum's basement, König uncovered a number of peculiar artifacts, including a clay jar five inches in length containing a copper cylinder surrounding an iron rod. The artifacts came from an excavation in Khujut Rabbou'a (a village just outside of Baghdad) and dated to around 200 B.C., a time when the warrior-like Parthians ruled the area. The markings, however, indicated that the objects were most likely created by the Sassanians, the last people to rule Middle Persia before the rise of Islam.[1] Testing revealed that the jar had once contained a corrosive agent, perhaps vinegar or wine that had soured.

In 1940, König published a controversial paper outlining his belief that the jar was, in fact, the earliest battery ever discovered and that it provided evidence that the ancients may have known about electro-plating, the process of using an electrical current to create a noncorrosive coating on various metals. Panned at the time, König's theory received little attention until 1993, when Paul T. Keyser, a researcher at the University of Alberta in Edmonton, published a competing theory in the *Journal of Near Eastern Studies*.[2] According to Keyser, the jar (and similar ones uncovered since 1938) was used by the people of Mesopotamia much like the Greco-Romans used elec-tric fish, to generate a mild, anesthetizing electrical shock. He theorized that the Sassanians may have stumbled upon the effect when using a bronze spoon in an iron bowl containing vine-gar or some other acidic liquid.[3] Anyone whose lips or hands touched both the bowl and the spoon at the same time would experi-ence a tingling sensation.

While Keyser's spec-ulations have stimulated many arguments among archaeologists, they stem in large part from the fact that the "Baghdad batter-ies" so closely resemble the first modern-day batteries, known either as Galvanic or Voltaic cells (depend-ing on whether one wants to honor Luigi Galvani, the father of modern electrochemistry—who also inspired Mary Shelley's *Frankenstein*—or Alessandro Volta, who relied on Galvani's work to create the first practical elec-trochemical cell). These early batteries generated electricity from the spon-taneous redox reaction between two different metals suspended in a salty solution. One metal, called the "anode," loses elec-trons (oxidation), while the other, called the "cathode," attracts them (reduction). As the electrons move from anode to cathode, a wire may be suspended in the solution to conduct the resulting A

TOP: Luigi Galvani
ABOVE : Alessandro Volta
BELOW: Baghdad Battery

when rubbed together, redox can occur spontaneously when certain substances are brought together. In fact, spontaneous redox is what powers most batteries.

Understanding Electromagnetism

Stimulating the movement of electrons through a chemical reaction may have been the first—and simplest—way to generate an electrical charge, but it is not the only way. In fact, our modern electrical system depends on a completely different method, one that uses magnets instead of chemicals. Most people understand how magnets work, by attracting oppositely charged objects to their poles. At one time or another, most of us have observed iron filings line up along the magnetic field lines at the ends of a bar magnet or have used them to give Wooly Willy a mohawk and a moustache. But few people understand *why* magnets work. And, truth be told, there is quite a bit of debate about whether modern science understands magnetism at all.[4] But there is some consensus around a possible explanation. Without getting too technical, one answer has to do with the natural alignment of tiny electrically charged particles within an atom. These particles tend to spin in one direction or another, creating very tiny pushes and pulls on the matter within the atom. When enough of these particles start spinning in a single

direction, the shared alignment generates a force of attraction that we know as magnetism. From this explanation (and some limited experience with refrigerator magnets or magnetic toys), one could think that magnetism is relatively weak. But physicists have learned that magnetism is an extremely strong force, vastly stronger even than the force of gravity.

When matter is charged, but the charge is stationary (that is, it stays within the object that is charged), it creates an electric field. We can say that the matter has taken on a physical property, like density or color. In simple terms, an electrical field describes the *intensity* of electrical charge, or its concentration relative to the

Wooly Willy

François Arago: The Swashbuckling Father of Electromagnetism

The discovery of alternating current is owed to a series of fortuitous events in the amazing life of the French scientist François Arago. Arago was born in the Pyrenees Mountains to a very influential political family. His father, in fact, was the treasurer of the mint for the region. After dropping out of the École Polytechnique, France's most renowned public institution for scientific education, Arago was appointed secretary of the Paris Observatory and was commissioned to complete measurements of the earth's meridian arcs. These highly accurate determinations of the distance of the globe's longitudinal lines were essential to astronomy and marine navigation.

In order to carry out his task, Arago traversed the ridgeline of the Pyrenees mountains, lighting fires as far south as the Balearic Islands and as high as the top of Mount Galatzó in the Catalan region. Not surprisingly, the Spanish were highly suspicious of a well-connected French patriot carting sophisticated measuring instruments along the Spanish border. Though he attempted to disguise himself as a Spaniard from Mallorca, Arago's true identity was soon uncovered. Branded a spy for the French army (which, incidentally, was preparing to invade Spain), he was quickly captured and imprisoned on Mallorca in the circular fortress of Bellver castle.[5]

Somehow, Arago managed to convince the warden of Bellver that he was, in fact, a scientist. Rather than simply free him, however (which may have raised the ire of his superiors), the warden gave him an opportunity to escape. Carrying logs of the measurements he had managed to record prior to his capture, on July 29, 1808, Arago fled the castle and snuck aboard a fishing boat heading to Algiers. There, the French consul supplied him with a forged Austrian passport and put him on a boat to Marseille, France. Unfortunately, fate would not have it. Once in open water, a Spanish warship boarded the boat, recaptured Arago, and imprisoned him in the Catalan port city of Rosas.

Once again Arago managed to convince the authorities of his true intentions and they quickly put him on a boat, again heading for Marseille. But a black cloud followed Arago—quite literally. A storm raging in the Mediterranean blew the boat all the way to the coast of North Africa, where Arago was captured by Muslim pirates. The wily Frenchman, however, convinced his captors that he wanted to convert to Islam. As a result of his "newfound faith," Arago was permitted to return to Algiers, where the French consul put him on yet another ship bound for Marseille. Fortunately, the third time was the charm. Arago finally reached French soil on July 2, 1809, nearly a year after he escaped Bellver and first set out for home.

François Arago

Michael Faraday: The Laws of Electromagnetic Induction

Michael Faraday's discovery of the laws of electromagnetic induction was as much due to luck as to the culmination of years of effort. Although he would go on to become one of the greatest luminaries in physics, Faraday came from humble beginnings. His father, James Faraday, worked as a blacksmith in Yorkshire in the north of England. In 1791, James Faraday moved his family to Newington Butts, a small village just outside of London, where he hoped work would be more plentiful. Shortly thereafter, Michael was born. Despite his efforts (and partially due to his poor health), the elder Faraday failed to find steady work and the impoverished family moved from one squalid neighborhood to another as young Michael and his siblings tried as best they could to gain an elementary education. By 1805, when he was thirteen years old, Michael was expected to start contributing to the family finances. He was fortunate enough to be taken

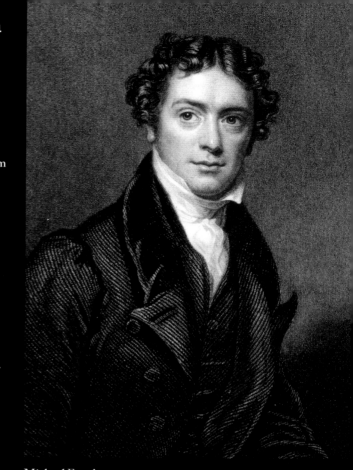

Michael Faraday

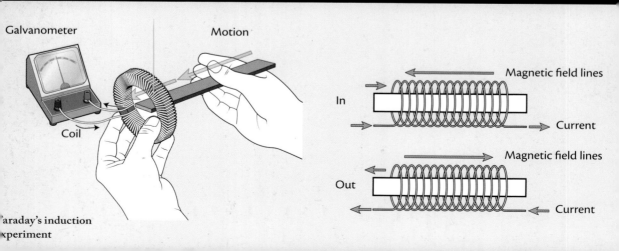

Faraday's induction experiment

Humphry Davy

Davy, an esteemed chemist and inventor, who lectured on a relatively new scientific concept called galvanism (known today as electrochemistry), the generation of electricity through a chemical reaction. When grade school students today create makeshift batteries by placing copper and zinc electrodes into a potato, they are performing a kind of galvanism that would have appeared quite impressive in the early nineteenth century. It certainly appeared impressive to young Faraday, who became enthralled with the concept of electricity.

After completing his apprenticeship, Faraday resolved to become a scientist. He had little formal education and a family of no notoriety. What he did have, however, was extraordinary pluck. So he set about writing to Davy, sending the chemist detailed notes he had taken while listening to Davy's lectures at the Royal Institution. In 1813, Davy fired his personal assistant, apparently for getting into a fistfight (though the historical records are a bit sketchy). Impressed with the bookbinder's ambition and perseverance, Davy offered Faraday the job, which the enterprising young bookbinder quickly accepted.

In October of 1813, Davy set out on a scientific tour of Europe, dragging the newly hired Faraday with him as his assistant and secretary. For eighteen months, Davy and Faraday bounced from one European capital to another, meeting André-Marie Ampère (the French physicist and founder of classical electromagnetism) in Paris and Alessandro Volta (the Italian inventor of the modern electrochemical battery) in Milan. Along the way, Faraday was exposed, directly and indirectly, to the thoughts and experiments of dozens of the best scientific thinkers of his time. This rich diversity of thought provided the foundation that would later lead Faraday to "discover" the innovations that would immortalize him in the annals of scientific history.

on as an apprentice bookbinder working for George Riebau, a successful London bookseller.[6] Faraday, however, was not satisfied with simply binding books; he read them as well.

During the seven years he served as Riebau's apprentice, Faraday befriended a medical student, J. Huxtable, and a clerk, Benjamin Abbott, both from more accomplished (and wealthier) families than Faraday's. Such company afforded him the opportunity to attend lectures at the Royal Institution, a London charitable organization devoted to scientific education and research. There, he was enthralled by Humphry

surrounding matter. The electric field tells us how strong an electrical attraction or repulsion another charge would experience at a particular location. All this is what we mean when we are talking about static charge; charge that is not moving.

When an electrical charge *moves*, from one piece of matter to another (say, between the electrons in the atoms of a conductive metal), however, the movement of the charge creates an entirely different field: a magnetic field. The movement could be on the large scale– say, if a charged particle moves through battery fluid–or it could be on the micro-scale, even as electrons remain local to their respective atoms and molecules, but engage in motions we might visualize as "spinning" or "circulating." Don't be fooled, though. Even as a charge moves from one piece of matter to another, if we were to freeze the charge at any given time—like interrupting a song during a game of musical chairs—the charge would still be contained within whichever particle held it when we stopped. Thus, as an

electrical charge moves, it generates *both* an electric field (from the mere *presence* of charge) *and* a magnetic field (from the *movement* of the charge). Unsurprisingly, we call this dual phenomenon an "electromagnetic field."

Understanding the relationship between the electrical properties and magnetic properties of matter took dozens of scientists hundreds of years to figure out. In fact, the electromagnetic properties of charged particles were little understood until the British mathematician and physicist Michael Faraday discovered the laws of electromagnetic induction in 1831.

Faraday's discoveries, like most innovations, did not spring from thin air. Rather they were deduced from bits of existing information gathered by scientists who may not have understood their significance at the time. In 1820, for example, a little-known French chemist named François Arago made a series of experiments that would prove pivotal to Faraday's understanding of electromagnetism. In one experiment, Arago suspended a

Arago's Wheel

magnetized needle above a spinning copper disc (a configuration that would come to be known as "Arago's wheel") and discovered that the needle would begin to rotate the same direction as the disc. However, if the needle was fixed, it would somehow affect the copper disc, slowing its rotation. Arago knew there was some force acting between the magnetic needle and the copper wheel, but he could not explain it in scientific terms.

Although Arago's observations hardly sound like a monumental breakthrough, when he presented these and other findings to members of the French Academy of Sciences (including the famous French physicist André-Marie Ampère), he helped rouse a collective effort among Europe's premier thinkers to develop a mathematical and physical theory to understand the relationship between electricity and magnetism.

Around 1810, shortly after a young François Arago returned from several years in captivity abroad, British engineer Charles Babbage (best known for inventing the first mechanical computer) arrived at Cambridge University and quickly formed a friendship with John Herschel, son of the British astronomer William Herschel (famous for his discovery of the planet Uranus in 1781).§ Like many recent grads today, upon receiving their diplomas, Babbage and Herschel lacked direction. Being from notable (and wealthy) British families, the two decided to take some time and travel abroad. In France, they visited the Society of Arcueil, an informal circle of some of the

best French scientists that met regularly in the Paris suburbs from 1806 until about 1822. Among the Society's members was none other than the adventurous François Arago.

From their time among the Society, Babbage and Herschel no doubt learned of Arago's experiments, for upon returning to England, they worked together to demonstrate the reverse of Arago's spinning copper disc trick. That is, rather than spin the copper disc, they wondered what would happen if they spun the magnet instead. They modified Arago's experiment by placing a spinning horseshoe magnet beneath a copper disc suspended by a wire. Although nothing appeared to attach the disc to the magnet, it nonetheless rotated in the same direction as the magnet. They surmised that there must be some relationship between the magnet and the disc, though, at the time (like Arago) they could not quite explain it.

Babbage and Herschel's experiments piqued the interest of Humphry Davy, who began to undertake his own experiments with electricity. Given his background in chemistry, however, Davy was far more interested in the electrical properties of certain chemical elements and less concerned with the relationship between magnetism and electricity. Consequently, rather than explore how magnetism might generate mechanical energy, he embarked on a series of experiments converting electrochemical to mechanical energy (ultimately creating the first battery-powered DC engine sometime around 1821).

Faraday's curiosity, however, stretched beyond electrochemistry. Instead of converting electrochemical to mechanical energy, he went the opposite direction. Working from Arago's, Babbage's, and Herschel's configurations, he connected a copper wire to the positive terminal of a battery, wrapped the remaining wire around one side of an iron ring, and left the end dangling near the negative terminal of the battery. He wrapped another copper wire around the opposite side of the iron ring and connected both ends to a galvanometer, an early device used for detecting an electrical charge. When Faraday closed the circuit on one side of the ring by connecting the loose copper wire to the negative terminal of the battery, the galvanometer detected an electrical current in the copper wire wrapped around the other side of the ring.

In theory, the charged wire that wrapped around the left side of the ring should have created a closed circuit when connected to both terminals of the battery. But Faraday observed an extra wave of electricity in the right side of the ring that he surmised must have been "induced" by his electrification of the left side.

Faraday tried a different configuration. This time he wrapped the copper wire all the way around the iron ring and attached both ends only to the galvanometer. Then he passed a bar magnet through the ring, observing that the needle of the galvanometer jumped only when he moved the magnet. If he held the magnet motionless through the ring, the needle did not move. Faraday had, in essence, found an explanation for the effects that Arago, Babbage, and Herschel had documented. He realized that the movements those scientists had observed were the product of the movement of a magnetic field near a conductor. Faraday supposed that "lines of force" (what we know today as a magnetic field) emanated from the poles of a magnet and "induced" an electromagnetic force in the conductor. However, the strength of the force depended

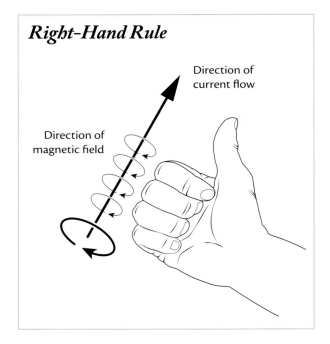

Right–Hand Rule

Direction of current flow

Direction of magnetic field

on how fast the magnet moved (in any direction) near the conductor. He expressed the relationship in what would become known as *Faraday's Law of Induction*: the strength of the electromagnetic force induced in any closed circuit is equal to the amount of change one causes in these lines of magnetic force.¶

Scientists had known for many years that every electric current was surrounded by a magnetic field.** What Faraday discovered was, essentially, that the opposite could be true as well: a magnetic field can generate an electric current—but only under certain circumstances. First, the magnetic field had

---

¶   Rather than "lines of force," scientists today refer to "magnetic flux," a measurement of the amount of magnetism that passes through a static surface at any given time.

**   An electric current passing through a conductor like a metal rod or wire produces a magnetic field in a circular pattern around the conductor. Scientists have developed the "Right-Hand Rule" as a handy way of describing this relationship. If you give the "thumbs up" sign with your right hand, the thumb corresponds to the direction of the initial electrical current and your fingers indicate the direction of the magnetic field that curls around it.

to move, fluctuate, or change in some way. Second, a conductive material (like a copper wire) had to be present in which electric charges were free to move in response to this fluctuating field. If these conditions were met, according to Faraday, an initial electric current (that creates a magnetic field) would generate a secondary electric current when the magnetic field was intercepted by a conductive material. It was not that this phenomenon—known as "electromagnetic induction"—diverted the initial current. Instead, induction created a whole new current. And the new current, of course, would create a new magnetic field, which, if intercepted by a conductor, would generate a new current, and so on. The effect is an endless feedback loop of electrical currents and magnetic fields, "the electromagnetic equivalent of nesting Matryoshka dolls."[7] Essentially, Faraday had discovered how to generate electricity using magnets, a phenomenon that would prove critical to the innovation—the polyphase AC motor—for which Tesla is best known.[††]

Understanding Direct Current (DC) and Alternating Current (AC)

The charge in a battery only flows in one direction, from the anode to the cathode. Batteries, therefore, will only generate a direct current (DC). Our modern electrical system, however, relies less on electrochemical energy from batteries than on turbines, which induce

electrical currents by spinning magnets. This method, however, has an interesting side effect: it can generate both direct and alternating current (AC). Unfortunately, most explanations of the difference between DC power and AC power—even those found in basic electrical engineering textbooks—simply restate the definition inherent in the name: a direct current moves in one, direct current; by contrast, an alternating current changes direction (alternates) at times corresponding to a frequency. Unsurprisingly, these explanations provide little help to a layperson seeking to truly understand the difference between the two types of power (and even less help understanding why the difference even matters).

If an electrical charge is a physical property of a material, an electrical current is the manner by which that property moves from one place to another. In the case of electromagnetic induction, it is like a tidal wave, transmitting energy across a vast ocean of atoms without actually moving the matter very much.

Like a tidal wave, a wave of electromagnetic energy can take different forms. Imagine a Newton's cradle, one of those toy pendulums often found on the desks of stodgy investment bankers. It usually consists of five or six identical steel balls, each touching its neighbor, each suspended by two strings connected to two parallel bars. It is easy to imagine an investment banker swinging one of the end spheres outward and then letting it drop toward the others. We know what will happen. The sphere swings down and hits the others. But only the sphere at the other end moves. The energy of the first sphere is transferred through the series of spheres until it forces the sphere at

---

†† An important thing to keep in mind is that these feedback loops do not make a perpetual motion machine—they do not create something out of nothing. To the extent that any work is to be extracted from the induced current, an equal or greater amount of work has to be invested at the beginning of the process to sustain the fluctuation of the magnetic field. This will come from the work it takes to make the generator turn, as we shall see very soon.

the other end to swing outward. Once started, this energy is transferred back and forth, each end sphere swinging out and crashing back into the line of spheres in the middle. While the end spheres swing back and forth, none of spheres in the middle ever appears to move.‡‡

We are used to thinking of kinetic energy as the energy of motion. Yet, with the desk pendulum, we can watch kinetic energy transferred between steel balls that appear motionless. In reality, the kinetic energy is converted to potential energy through the compression of the metal on one side of the sphere. This energy is then released again as kinetic energy when the metal decompresses. This compression and decompression, however, is so slight and happens so quickly that most people don't even perceive it. This action is, in fact, more analogous to the way an impulse of electrical energy travels through charged particles than more conventional illustrations that describe electricity as a flow of electrons. Electrons absorb an electrical impulse and then pass it on, sometimes at nearly the speed of light and other times quite slowly (depending on the nature of the conductor through which it is passing).

In any event, it is not necessary to understand Newton's laws of motion to make use of his cradle to consider how kinetic and potential energy affects the difference between AC power and DC power. Imagine an electrical conductor consists not of a wire, but of a string of identical steel spheres, each barely touching

Photo of a Newton's Cradle

its neighbor. (After all, at the atomic level, a metal wire is nothing more than a string of spherical atoms, each barely touching one another.) To transmit an electric current we are faced with a challenge: how can we apply force at one end of the string to move the metal sphere at the other end?

There are (at least) two possible solutions. We can push on the sphere at one end and expect that the effort we expend will work its way through the spheres, forcing the sphere at the other end to move the same amount. This is much like direct current (DC) in an electrical circuit. The positive terminal repulses negatively charged electrons, creating an electrical charge that "pushes" on each subsequent electron, moving in one direction, from electron to electron, in a continuous, even flow.

The second solution is to shake the sphere at one end and vibrate it back and forth. As it makes contact with the neighboring sphere, it too begins to vibrate. As that neighboring sphere vibrates, some of its energy transfers to the next sphere, which begins to vibrate, and so on until all the spheres are crashing back and forth, absorbing and distributing energy from one another. Eventually, the energy we are expending solely by moving the first sphere within a defined space

---

‡‡ What we are witnessing is the conservation of momentum, a fundamental law of motion articulated by physicist Sir Isaac Newton. It is not, however, a perfect example of the law, which is why we can discern slight movements of the inner balls if we look closely.

Direct Current

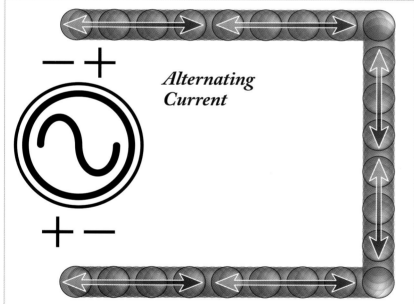

Alternating Current

At first glance this solution might appear like a chaotic game of atomic red-rover. In practice, however, it provides some flexibility that proves advantageous, especially in transmitting electricity over long distances.

Transmitting Electricity

Despite the gory theatrics of public electrocutions, the "War of the Currents" was more about electrical transmission than safety. Early electrical motors, which almost universally ran on DC, were incompatible with an electrical distribution system that relied on transmitting power from large, central generating stations to customers more than a few kilometers away. DC power (at least at voltages available in the late nineteenth century) tended to dissipate very quickly the farther it had to travel through a conductor. When trying to distribute DC power over wide distances, therefore, early electrical engineers faced two choices, neither of which was very practical: either build lots of generating stations (almost all of which would burn dirty coal) close to the customers that needed the electricity, or transmit DC power over thick-enough wires (almost

will find its way through the entire string.

By alternating the position of positively charged and negatively charged terminals, an AC circuit "vibrates" the electrons within the conductor, causing them to oscillate back and forth, passing energy to neighboring electrons.

all of which were made of expensive copper), so that there was enough electricity left at the other end to be of any use. Not only were thick copper power lines bulky and expensive, there was a limit on just how thick they could be made. For all practical purposes (and despite Edison's assertions otherwise), DC power could not be relied upon to supply electricity to a nation expanding westward at an alarming pace.

Understanding Voltage

The advantage of using alternating current instead of direct current for long-distance transmission is largely due to voltage, one of the most difficult concepts of electricity to understand (and also to explain). When an attractive force exists, it automatically creates potential energy in everything it attracts. Think of the attractive force of the Earth's gravity. We can calculate the potential energy of an object by how far it is lifted off the Earth's surface. Its energy is potential because, at some time, the object could fall toward the Earth, accelerated by the Earth's gravity. This active form of energy can then be used to perform work. In a DC circuit, this charge moves in a single direction, changing the potential energy of each electron as the charge changes its position.

What makes voltage so confusing to many is that it is a measure of the *difference* in the potential energy along an electrical circuit. It is a relative measure. Put more simply, if current is the measure of "how much stuff is going," voltage is a measure of "how badly the stuff wants to get there."[8] In mechanical terms, voltage is often compared to the pressure of water in a pipe. In fact, novice engineers often are told to think of electrical circuits like a closed plumbing system.

Where current is like the actual volume of water contained within the system, voltage is equivalent to the water pressure in any given pipe.[§§] This analogy is not exactly correct. Voltage would be more analogous, for example, to the height difference between a reservoir of water and the downhill end of a pipe connected to it. Yes, water in the reservoir would be at a different pressure than water spurting out of the downhill end of the pipe. But assume there is a drought and the reservoir is dry as a bone. Voltage would still describe the *difference* in potential energy between the reservoir and the pipe. We simply have to wait for water to fill the reservoir in order to make that difference apparent.

The measure of the difference in potential energy becomes essential when thinking about the difference between DC and AC. Because the positive and negative terminals of a DC circuit are static, voltage must remain constant. Each electron in the circuit has measurable potential energy due to its distance from the terminals. However, at one point or another, as an electrical charge flows through a string of electrons in a DC circuit, the charge will occupy every position along the circuit. As the charge moves in one direction, its potential energy shifts (from higher to lower, or lower to higher) always in the same proportion, like a well-trained army marching in step. While the potential energy of soldiers at each electron changes, the difference in potential energy between the soldiers at any two positions is always the same. They always have the same desire to get to a terminal, and all, therefore,

---

§§ Before the concept of voltage was fairly understood, early electrical engineers (Tesla included) often referred to putting current "under pressure."

march toward the terminals at the same speed. The charge moves, but always by the same, predictable amount toward or away from the stationary terminals. This stability is why batteries, which have fixed positive and negative terminals, may come in different voltages (1.5 volts, 6 volts, 9 volts, etc.), but always supply only the labeled voltage. To change voltages, you have to change batteries.

At first this relationship might seem completely academic. But it becomes supremely important when thinking about the best way to transmit electrical charges over long distances. It turns out that one of the best ways to limit the amount of electricity dissipated over long-distance transmission lines is to find a way to package the same amount of energy in less current. This is where voltage plays a critical role—specifically, the quirky ability to change voltage in an AC circuit. To understand this phenomenon, it is handy to think of electricity as a flow of electrons.

The way energy is transmitted through a circuit is the product of its current (say, the raw number of electrons passing through the line) and its voltage (the difference in potential energy between each electron). If DC power is like a well-trained line of soldiers, then the

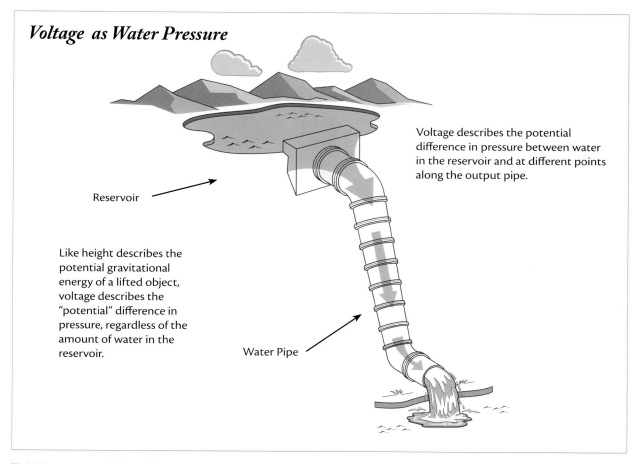

Voltage as Water Pressure

Reservoir

Voltage describes the potential difference in pressure between water in the reservoir and at different points along the output pipe.

Like height describes the potential gravitational energy of a lifted object, voltage describes the "potential" difference in pressure, regardless of the amount of water in the reservoir.

Water Pipe

Uphill reservoir with downhill pipe

strength of an army of DC soldiers depends on how many electrons are marching in each row (current) and how fast the rows are marching in step (voltage).

It is important not to confuse either current (the flow rate of electrons) or voltage (the "pressure" applied to the electrons) with actual power. Power, the product of current and voltage, is a measure of the rate at which the energy within a moving electrical charge can be converted into some other form that can perform work. For example, electrical energy can be converted directly to mechanical energy to perform some work (as in turning a crankshaft) or indirectly to potential energy that can be stored (as in compressing air to generate pressure that may be released later). Power, in other words, requires both current and voltage to "work."

Ohm's Law and the Advantage of Alternating Current (AC)

In the early nineteenth century, a discontented German mathematics teacher named Georg Ohm thought he could make more money as a physics lecturer. The son of a locksmith, Ohm had some practical knowledge of mechanical devices and began tinkering with the equipment in the physics laboratory of Jesuit Gymnasium in Cologne, where he taught. Mostly, this consisted of playing with the electrochemical cells recently invented by the Italian scientists Luigi Galvani and Alessandro Volta.

Unlike his contemporaries, Ohm rejected the idea that electricity amounted to a force that acted on objects at a distance. Rather, he believed that electricity was a force transmitted between "contiguous particles," much like

the string of steel spheres in a Newton's cradle. By 1827, he published an article entitled "The Galvanic Circuit Investigated Mathematically" (Ohm apparently favored Luigi over Alessandro). In it, he articulated a mathematical relationship between an electrical current and its voltage when applied across a conductor. The resulting equation became known as Ohm's Law and it is the cornerstone of today's modern electricity grid.

Ohm's Law:

I *(current)* $= V$ *(voltage)* $\div R$ *(resistance) or*
V *(voltage)* $= I$ *(current)* $\times R$ *(resistance)*

Don't be daunted by the mathematics. The relationship is really quite simple. If we assume that the resistance in an electrical wire remains constant, then there is a proportional relationship between current and voltage. That is, if you increase the voltage of a charge, you also increase the current. Likewise, if you decrease voltage, you decrease current.

The proportional relationship between voltage and current becomes particularly useful when thinking about how to transmit power. In essence, it means the two are interchangeable, like trading four (relatively heavy) quarters for one (relatively light) dollar bill. Like money, power—the product of both current and voltage—can be packaged in either form.

If you are still confused, let's return to our analogy of an electrical charge as a marching army of electrons. In a DC circuit, the army must march in lockstep at whatever speed its regiment started. Soldiers are given their orders, sent on their way, and are forced to maintain formation for however far they march. Soldiers in an AC circuit are more easily directed (if no less regimented); they have the option of receiving new orders mid-march. At any time, from battlefield to battlefield, soldiers in the AC army can be directed to start marching at a different speed. (But, as we shall see, new orders come at a price.)

Resistance and Line Loss

As you can imagine, the ability to direct the speed of an army of electrons mid-march can be quite useful, especially when considering the terrain over which they must march to get from one place to the next. As an electrical charge moves through a conductor, it meets resistance: literally the measure of the conductor's opposition to the passage of the charge. In most cases, resistance creates a kind of atomic friction, which generates heat. The greater the power that wants to pass

through the conductor, the more resistance and, hence, the more heat is produced. If the conductor is made of metal (as most electrical wires are), the heat will cause the conductor to expand. It is this thermal expansion, in fact, that causes power lines to sag, come into contact with tree branches, cause short-circuits, and trigger widespread power outages.

Resistance not only creates problems by heating power lines, the heating also uses up some of the power that is being transmitted through the line (in some cases, a great deal of the power). While almost all conductors will put up some resistance to power that wants to flow through them, the amount of resistance varies depending on the *form* of the power.

It turns out that a conductive wire will "leak" power at a faster rate than it will "leak" voltage. Remember that voltage is a relative

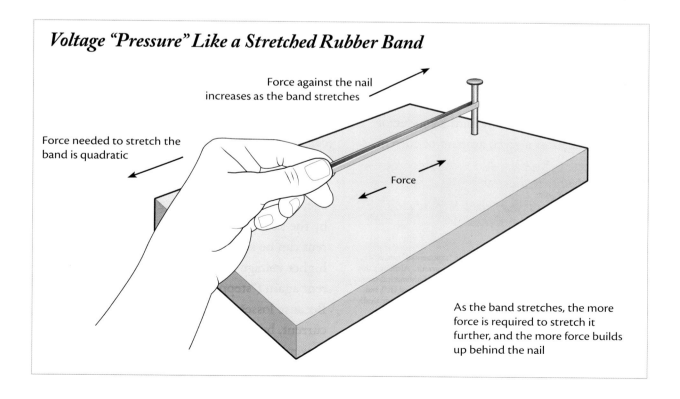

Voltage "Pressure" Like a Stretched Rubber Band

Force against the nail increases as the band stretches

Force needed to stretch the band is quadratic

Force

As the band stretches, the more force is required to stretch it further, and the more force builds up behind the nail

measure, like the difference in potential energy between a reservoir of water stored up a hill and the potential energy of water at the end of a downhill pipe. The drop in the potential energy of water as it flows downhill through the pipe is constant and depends upon the length of the pipe. Voltage drop along a conductive wire (like a transmission line) acts the same way. The line will lose voltage at a rate that is constant and linear—the longer the line, the greater the loss.

On the other hand, a transmission line will "leak" power at a quadratic rate, a rate that increases with the square of the current.¶¶ This is due in part to the fact that, like a stretched rubber band, a conductor's resistance will increase the more current tries to "push" through it. Also like a stretched rubber band, the more current you try to push through a conductor, the harder it will push back. The rate at which this resistance will increase, however, is "quadratic" not linear. This is due to the relationship among current, voltage, and resistance articulated in Ohm's Law.

Using Ohm's law, scientists have derived a formula for calculating how hard a conductor will "push" against a given amount of current.

Power Lost to Resistance:

P (power lost) = I^2 (current) x R (resistance) ***

¶¶ Most people would say that power is lost at an "exponential" rate, since power loss increases with the square (exponent) of the current. Although this is the common terminology, it is technically incorrect. In mathematical terms, an "exponential rate" would be equivalent not to resistance squared (R^2), but to resistance raised to a changing exponent (R^N). Still, for the mathematically challenged, it may be simpler to think of the linear loss of voltage along a transmission line versus the "exponential" loss of current along the same line.

*** Mathematically, this equation can be derived entirely from the definition of power [P(power) = I(current) x V(voltage)] and Ohm's Law [V(voltage) = I(current) x R(resistance)]. Substituting I x R for V in the first equation renders: [P = I x (I x R)] or simplified to P = I^2R.

Just as Ohm's Law demonstrates that voltage has a proportional relationship to current (as one increases, so does the other), this formula demonstrates a similar proportional relationship between current and resistance. The difference, however, is important. While voltage and current are *directly* proportional, current and resistance are *quadratically* proportional. This difference becomes critical when determining how to transmit DC power (which has a constant voltage) over a long transmission line. Because the voltage of DC power cannot easily be changed using an electrical transformer, a given amount of DC power (especially in Tesla's time) generally only can be "packaged" and sent through a transmission line as current. Sending a large amount of DC power, therefore, requires sending a large amount of current (with quadratic line losses due to resistance).

However, with AC power, we have the option of "repackaging" the power as voltage (with *linear* losses due to resistance). Because we can alter voltage in an AC circuit, it is much like having the ability to direct an army to change speeds mid-march. Using electrical transformers (see Chapter Five), AC power can be repackaged as voltage ("stepped-up") at one end of a cable without losing any energy. Since power is the product of current and voltage, every increase in voltage decreases current by the same factor. In other words, a lower current can be sent through a transmission line as higher voltage and then repackaged as current again ("stepped-down") at the other end. Because losses vary with the square of the current, but linearly with voltage, by sending power through the conductor at higher voltage and lesser current, we can trade quadratic

The Life & Times

of

Nikola Tesla

Fall 1875 - Tesla enrolls at the Joanneum Polytechnic School in Graz (Austria)

1878 - Pavel Yablochkov demonstrates his AC arc-lighting "candles" at the Paris Exhibition

Fall 1878 - Impoverished and depressed, Tesla flees to Maribor (Slovenia)

1879 - Walter Baily presents his paper on rotating magnetic fields before the Physical Society of London

April 7, 1879 - Milutin Tesla dies of an unspecified illness at age 60

1861–1862 - James Clerk Maxwell presents a series of four papers on the nature of electromagnetism

1862 - Milutin Tesla moves his family to Gospić (Croatia)

1868 - Earliest date Mahlon Loomis may have demonstrated his kite-based system of wireless transmission (Washington, DC)

Nov. 24, 1831 - Michael Faraday presents his groundbreaking paper on electromagnetism before the British Royal Society

1832 - Hippolyte Pixii invents the first hand-cranked AC generator; André-Marie Ampère designs the commutator to convert the resulting AC to DC

1830

1860

1875

1855

1870

1855–1856 - James Clerk Maxwell presents his groundbreaking paper on "Faraday's Lines of Force" to the Cambridge Philosophical Society

1856 - Heinrich Geissler invents the first crude fluorescent tube

July 10, 1856 - Nikola Tesla is born in Smiljan, which was part of the Austrian Empire (now Croatia)

April 30, 1872 - William Henry Ward is granted a patent on his tower method of wireless telegraphy

July 30, 1872 - Mahlon Loomis is granted a patent on his method of wireless telegraphy

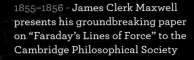

COLOR KEY:

YELLOW = Location of Nikola Tesla and significant life events

BLUE = Innovation of the polyphase AC motor

GREEN = Innovation of the transformer and Tesla coil

RED = Innovation of radio and wireless transmission

VIOLET = Chronology of patent disputes and court decisions

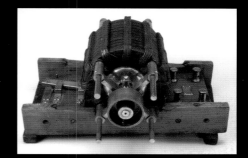

1887–1888 - Heinrich Hertz proves Maxwell's theories through a series of experiments

March 1888 - Tesla visits the Mather Electric Company to try to sell his AC motor, but does not mention the split-phase design (Hartford, CT)

Mar. 11, 1888 - Galileo Ferraris presents his paper describing a split-phase motor to the Royal Academy of Science in Turin

April 1888 - Tesla claims he first mentioned his split-phase design to his patent lawyer, James Paige

May 1, 1888 - Tesla is granted a patent on his polyphase AC "electro-magnetic" motor

May 15, 1888 - William Stanley visits Tesla's lab on Liberty Street, but claims Tesla did not mention or demonstrate the split-phase design

1885 - Zipernowsky, Bláthy & Déri first demonstrate the ZBD transformer in Budapest

Spring 1885 - Westinghouse purchases the rights to the Gaulard-Gibbs transformer design

Fall 1886 - Tesla moves into a laboratory at 89 Liberty Street (NYC)

Oct. 5, 1886 - Amos Dolbear is granted a patent on his wireless "mode of electric communication"

Summer 1888 - Tesla moves to Pittsburgh to work on the manufacture of his motor at Westinghouse's factory

Dec. 8, 1888 - Tesla files a patent application on the split-phase design for his AC motor

1889 - Tesla returns to New York City, moves into the Astor House at the corner of Broadway and Vesey Streets, and moves his laboratory to 975 Grand Street (NYC)

1889 - Westinghouse brings a patent infringement suit against New England Granite Company

1885

1880

Fall 1880 - Tesla moves to Prague (now in the Czech Republic) to attend the Karl-Ferdinand University (now Charles University), but never enrolls (Prague)

1881 - Marcel Deprez presents his paper on rotating magnetic fields to the French Academy of Sciences

January 1881 - Tesla moves to Budapest (Hungary), takes a job with the Hungarian Central Telegraph Office, and meets Anthony Szigeti (Budapest)

1882 - Tesla claims to have worked on a ring transformer he found while hanging around the Ganz Works (Budapest)

1882 - Amos Dolbear first demonstrates how ground currents could be used for wireless telegraphy

February 1882 - Tesla claims that the idea of a motor using a rotating magnetic field struck him during a sunset walk in Városliget City Park (Budapest)

April 1882 - Tesla and Szigeti move to Paris to work for the Continental Edison Company (Paris)

1883 - According to Szigeti, Tesla first discusses designs for an AC motor without a commutator

1883 - Lucien Gaulard & John Gibbs first demonstrate their transformer design in London

1884 - Lucien Gaulard & John Gibbs demonstrate long-distance transmission of AC power in Turin

June 6, 1884 - Tesla arrives in New York aboard the SS *City of Richmond* (Szigeti followed in May 1887)

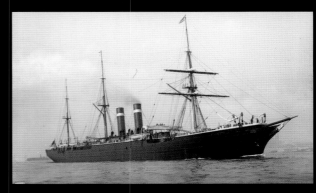

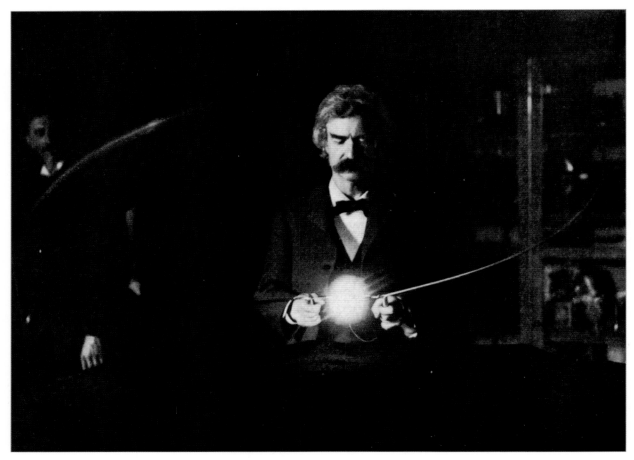

Mark Twain observing a spark gap in Tesla's Liberty Street laboratory

resistive losses for linear resistive losses, and save a huge amount of power in the process.

The ability to transmit AC power over longer distances with less energy lost to resistance is the primary reason that Westinghouse was winning the War of the Currents despite Edison's gruesome campaign to paint AC as more dangerous than DC. In the end, customers were less concerned about safety than they were about the practical ability to transmit AC power over long distances and at cheaper prices. It is also why Tesla's AC motor design became the standard that would govern the development of the modern electricity grid (and, eventually, lead the modern genius to be lauded as "inventor of the electrical age").

June 13, 1905 - U.S. Patent Office reverses itself and grants Marconi a patent for "wireless telegraphy"

1909 - Tesla rents offices in the Metropolitan Life Tower at East 23rd Street and Madison Ave. (NYC)

1913 - Tesla moves his offices to the Woolworth Building at 233 Broadway (NYC)

1930 - Tesla is evicted from the Hotel Pennsylvania and moves into the Hotel Governor Clinton at 371 7th Ave.

Jan. 2, 1934 - Tesla moves into his last residence, room 3327 at the New Yorker Hotel at 481 8th Ave. (NYC)

1905

1930

1915

1943

~01 - Judge Albert Thompson pholds Tesla's patent on the plit-phase motor in the Dayton an & Motor case

pt. 14, 1901 - President William cKinley is assassinated; Judge ohn R. Hazel (of the federal strict court for the southern strict of New York) swears in heodore Roosevelt as president

ec. 11, 1901 - Tesla breaks round on his wireless tower at ardenclyffe, Long Island, NY

ec. 12, 1901 - Marconi transmits gnal across the Atlantic; ough widely reported on, the ventor and his assistant actually tnessed the event.

1902 - Sixth Circuit Court of Appeals upholds Judge Thompson's decision in the Dayton Fan & Motor case; Theodore Roosevelt appoints Judge Townsend to the Second Circuit Court of Appeals

1903 - Judge Townsend overturns Tesla's patent on the split-phase design in the Catskill Illuminating & Power case

October 1903 - U.S. Patent Office rejects Marconi's radio patent application, citing Tesla's patent on wireless transmission

1904 - Judge Hazel reverses the Second Circuit's precedent and upholds Tesla's patent on the split-phase motor in the Mutual Life Insurance case

1916 - In debt and unable to pay the taxes on Wardenclyffe, Tesla moves his offices to 8 West 40th Street, and declares bankruptcy

July 4, 1917 - Wardenclyffe tower is dismantled and sold as scrap to pay Tesla's debt to the Waldorf–Astoria

1917–1926 - Tesla is evicted from the Waldorf–Astoria and variously moves to Chicago, Milwaukee, Boston and Philadelphia consulting for various manufacturing companies

1926 - Tesla returns to New York and moves into a room at the Hotel Pennsylvania at 401 7th Ave. (NYC)

Jan. 7, 1943 - Tesla dies alone in his room at the New Yorker Hotel (NYC)

June 21, 1943 - On a 5-3 decision (with one Justice recused), the U.S. Supreme Court rescinds Marconi's radio patent partly on the grounds that it was anticipated by Tesla's prior wireless patent

1895 - Westinghouse, J. P. Morgan and Thomas Edison form a patent pool including Tesla's patents on his original AC motor design

Mar. 17, 1895 - Tesla's laboratory at 35 South Fifth Avenue burns to the ground (NYC)

May 7, 1895 - Alexander Popov presents a paper outlining a design for a radio receiver before the Russian Physical and Chemical Society

1896 - Tesla opens a laboratory at 46-48 East Houston Street (NYC)

Nov. 16, 1896 - The Niagara Falls power plant comes online, transmitting AC

Sept. 2, 1897 - Tesla files a patent on a wireless "system of transmission of electrical energy"

1898 - Tesla's patents on the split-phase design are added to the Westinghouse-Morgan-Edison patent pool

January 1899 - Tesla moves into rooms at the Waldorf-Astoria (NYC)

May 18, 1899 - Tesla arrives in Colorado Springs to perform experiments in wireless telegraphy

July 28, 1899 - Tesla picks up a radio signal he believes to be from Mars

1895

890

1890 - Tesla recreates Hertz's experiments in electromagnetic waves in his Grand Street laboratory

March 26, 1890 - Tesla files the first in a series of patents on the "oscillating transformer" that was the basis of his Tesla coil

Summer 1890 - Tesla's companion and assistant Anthony Szigeti either disappears or dies of an unknown illness

February 4, 1891 - Tesla files a patent incorporating a capacitor into his Tesla coil design

May 16–Oct 19, 1891 - Charles Brown & Mikhail Dolivo-Dobrovolsky demonstrate long-distance transmission of AC power using the Gaulard-Gibbs transformer (Frankfurt)

May 20, 1891 - Tesla gives his famous lecture at Columbia University, demonstrating high-frequency induction and wireless lighting

July 30, 1891 - Tesla is sworn in as an American citizen

Jan. 26, 1892 - Tesla arrives in London to begin a European tour promoting his invention of the split-phase motor

Feb. 5, 1892 - John Ambrose Fleming invites Tesla to his laboratory at University College, presumably to demonstrate his high-voltage induction coil

April 1892 - Tesla travels to Gospić just in time to see his mother die on April 4th

Summer 1892 - Tesla visits the Ganz Works where Zipernowsky was overseeing construction of a large AC generator and high-voltage transmission system

July 1892 - Tesla visits Heinrich Hertz at his laboratory in Bonn, where the two discuss (and disagree about) the nature of electromagnetic waves

Aug.–Sept. 1892 - Tesla returns to New York and moves into the Hotel Gerlach at 29 West 27th Street (NYC)

March 1, 1893 - Tesla demonstrates wireless lighting in a lecture before the National Electric Light Association (St. Louis)

May 1, 1893 - Westinghouse & Tesla demonstrate AC power and lighting at the World's Columbia Exhibition (Chicago)

July 7, 1893 - Tesla files a patent introducing electrical resonance into his Tesla coil design

1900

Jan. 15, 1900 - Tesla returns to New York City (NYC)

Feb. 19, 1900 - Tesla files an amendment to his 1897 wireless transmission patent

Aug. 29, 1900 - Judge William Kneeland Townsend of the U.S. federal court in Connecticut upholds Tesla's patent on his initial AC motor design in the New England Granite case

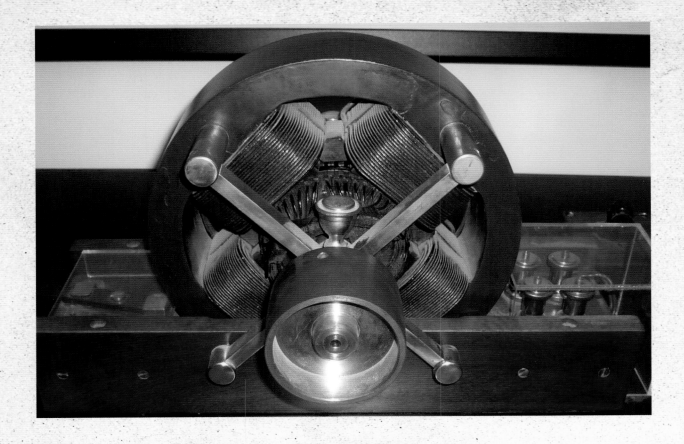

"*Great ideas are rarely the exclusive property of individual men. The history of the evolution of thought shows distinctly that novel ideas arise in certain generations simultaneously in the minds of some of the most talented men . . . and so it has been with the ideas and inventions that form the basis of the polyphase system.*"

—B. A. BEHREND, *ELECTRICAL WORLD AND ENGINEER*, MAY 6, 1905

chapter four

The Polyphase Alternating Current (AC) Motor

As you read this sentence there are an estimated seven hundred million electric motors of various sizes in operation around the globe.[1] Not only do electric motors power almost every aspect of modern living (from refrigerators to air conditioners to personal computers), they are also essential components in almost every industrial manufacturing process. Electric motors are critical to the operation of trains, subways, and other mass transit systems. As the cost of petroleum increased into the early 2010s, electric motors became increasingly vital to personal transportation as well. It is easy to forget that electric motors are crucial components of the supply chain that allows fossil fuel–powered motors to function. When Superstorm Sandy plunged the coasts of New Jersey and New York into darkness in 2012, thousands of drivers were

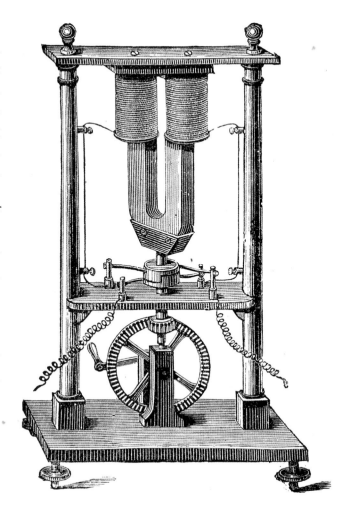

OPPOSITE: Historical photo of an early AC motor

RIGHT: Illustration of Hippolyte Pixii's hand-cranked AC generator

left stranded not because of a shortage of gasoline, but because almost all gas station pumps ran on electric motors.[2]

In addition to playing a vital role in the global economy, electric motors are central to the planet's environmental health. They are (by far) the single largest consumer of electricity on earth and account for 45 percent of all the power consumed worldwide.[3] If electric motors got first dibs every year on all of the electricity produced anywhere, the world would have to wait until June 16th before anyone could plug in anything else.

Small, relatively quiet, and often overlooked, these impressive tools are considered by some to be among the top ten discoveries of all time,[4] and by many to be the single greatest invention ever to spring from the mind of Nikola Tesla.[5] But the story of just who "discovered" the modern electric motor is more complex than popular wisdom (or Tesla's own accounts) would have us believe.

On October 12, 1887, Tesla filed an application in the United States Patent Office for an "Electro-Magnetic Motor."[6] He proposed a design that would convert a rotating electromagnetic field into mechanical energy to turn an axle. Certainly electric motors already existed and were even in widespread use, powering elevators, trolley cars, and even some industrial machinery by the time Tesla conceived of his version.* What made Tesla's design different enough to warrant a patent—and, incidentally, to win him notoriety as the father of the modern electrical age—were two innovations, which together revolutionized electrical engineering and paved the way for the bulk power grid that we still rely upon today. First, Tesla's design used a rotating magnetic field to convert an induced electrical current into mechanical energy. Second, it did this without requiring the use of a commutator. If none of this sounds impressive, don't despair. A bit of explanation about the basic engineering principles of an electric motor will help clarify why Tesla's famous design proved so groundbreaking.

First, it is important to understand that electric generators and electric motors are flip sides of the same coin. A generator is designed to convert mechanical motion into electricity using electromagnetic induction. A motor does just the opposite. It is designed to convert an electrical current (induced by electromagnetism) into the mechanical energy of motion.[7] In practice, the design of generators and motors is so similar that most generators can operate as motors and vice versa. As a result, innovations in the design of a motor usually also represent innovations in the design of a generator.

Evolution of the Electric Motor

Since the first electrical devices were powered by batteries, they necessarily had to be designed to utilize DC power. The electrical telegraph, for example, initially employed multiple wires—each representing a different character of the alphabet—with the ends submerged in glass tubes filled with acid. The sender applied an electrical current to a wire at one end. The current would travel through the wire and electrolyze the acid, releasing tiny bubbles of hydrogen in the tubes at the other

---

* We have already mentioned that Humphry Davy invented the first battery-powered electrical engine around 1821. The first working electric motor in the United States was built by Thomas Davenport, a Vermont blacksmith, in 1834, before anyone had even designed the first generator to power it. Using material he found around his house and blacksmith shop (including silk from his wife's wedding dress), Davenport constructed a model train powered entirely by electricity. See M.W. and E. Reilly, "Thomas Davenport (1802–1852): Inventor of the DC Electric Motor," Edison Tech Center Engineering Hall of Fame, 2010, http://www.edisontechcenter.org/DavenportThomas.html.

end. The receiver would watch which tubes bubbled, record the characters, and decipher the message.[8]

The first commercially available electrical motors in Europe were invented by British scientist William Sturgeon in 1832. Based on his observations of electromagnets, Sturgeon designed a machine that used DC power to run small machine tools and some printing presses. But powering them with electrochemical batteries proved so expensive that Sturgeon (and several copycat inventors) went bankrupt.[9] Partly as a result of these limitations, many inventors of the time turned their attentions toward electromagnetic induction as an alternative to battery power.

The simplest form of an electromagnetic motor or generator induces an electrical current by moving a wire relative to a magnet or moving a magnet relative to a wire. Sometimes moving the magnet inside a fixed armature turns out to be the more practical configuration, rather than moving the armature inside the stationary magnetic field. Thus, we can imagine the most basic generator as a bar magnet spinning within a loop of conductive wire.

If we picture magnetic force emanating from the poles of the magnet, it is easy to see that none of the lines of force intersect the loop of wire when the magnet is positioned parallel to the plane of the loop. As the magnet rotates, however, more of the imagined magnetic field lines intersect the wire loop. As the magnet continues to rotate, the more the imagined field lines intersect the plane of the wire, until the magnet is perpendicular to the plane of the loop (see figure above). With even further rotation, intersection of the plane

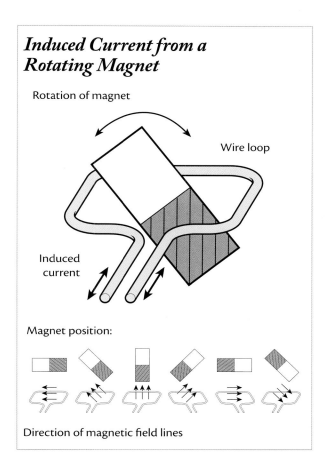

Induced Current from a Rotating Magnet

Rotation of magnet

Wire loop

Induced current

Magnet position:

Direction of magnetic field lines

reduces until the magnet is again parallel to the loop of wire. Thus, during one complete rotation, the amount of intersection of the plane of the loop increases, reaches maximum, and then decreases until the magnet reaches its original starting point.

Since an electromagnetic current is induced by the *rate of change* in the position of the magnet relative to the position of the wire, we can express the amount of induction as a sine wave. For half of the cycle of the wave, while the magnet's rotation increasingly intersects the plane of the loop, the rate of change is positive. That is, the wave approaches its peak. For the other half of the cycle, while the magnet's rotation returns to parallel, the

Basic Components of an Electric Motor

Although several versions of electric motors eventually achieved commercial success, almost all employed the same basic structure, including a stationary part (called a stator), and a rotating part (called a rotor). In the most common motor design, the stator consists of a field magnet with a constant magnetic field, and the rotor is comprised of coils of conductive wire (typically copper or aluminum) and an axle. The conductors that carry the current involved in the transfer of electrical work are also called the armature windings. Commonly, the rotor is the armature. In DC motors or generators, a commutator—used to convert between alternating and direct current—is usually attached to the axle and makes periodic contact with copper brushes attached to the stator.

Stator (Field Magnet)

The stator is the stationary part of an electric motor. In Tesla's design, it acts as a field magnet, periodically attracting or repelling the poles of the electrified, spinning armature.

Armature

The armature is the part of an electrical machine around which conductive coils are wound. In this case, it is the moving part or rotor. The armature carries current, which itself produces a magnetic field. The interaction between the rotor and stator magnetic fields— much like the attractive or repulsive force between two refrigerator magnets— creates a torque that can be used to perform work.

Commutator

A simple commutator is a pair of plates attached to the axle on the armature. The plates connect the wound coils of the armature with the source of the electromagnetism. As the armature spins, the plates periodically make contact with brushes (usually of copper wire or carbon) attached to the stator. Every time the plates reconnect with the brushes, they reverse the flow of the current. Commutators are only needed for DC machines.

Inside an Electric Motor

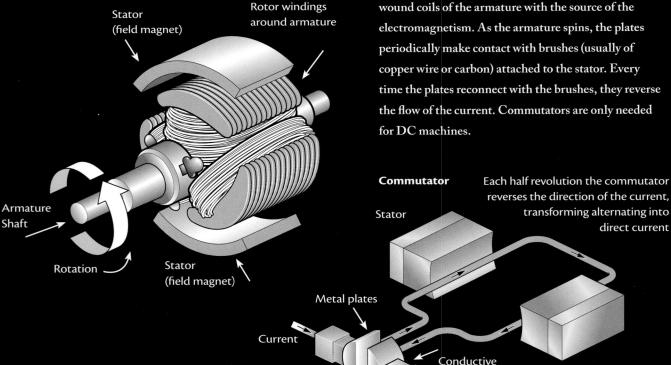

Stator (field magnet)

Rotor windings around armature

Armature Shaft

Rotation

Stator (field magnet)

Commutator

Stator

Each half revolution the commutator reverses the direction of the current, transforming alternating into direct current

Metal plates

Current

Conductive Brushes

rate of change is negative. That is, the wave diminishes. This fluctuation is what produces an alternating current. The frequency of the fluctuation corresponds with the number of rotations of the magnet over a given period of time. Thus, the rotation of the magnet is said to be "in phase" with the sine wave representing the alternation of the induced current.

DC circuits generally operate with two wires. One carries current from the generator (say, a battery) to whatever device is utilizing the energy (say, a motor). The other wire carries current from the motor back to the generator. However, after Michael Faraday published the results of his experiments in a groundbreaking paper presented before the British Royal Society in 1831, suddenly every scientist and amateur inventor wanted to experiment with induced electromagnetic currents.[10] When the first induction generators were invented (using rotating magnets), scientists faced the challenge of working with an electrical current that reversed polarity with every full rotation of the magnet. The alternating current was impractical for powering motors since, rather than a steady flow of electricity, the flow would pulsate with the rotation of the magnet inducing it. Electrical engineers struggled with how to induce a steady current from a rotary motion. In effect, they were seeking to do precisely what Professor Poeschl told Tesla could not be done (at least not without using a commutator to convert the pulses of an alternating current into the steady flow of a direct current).

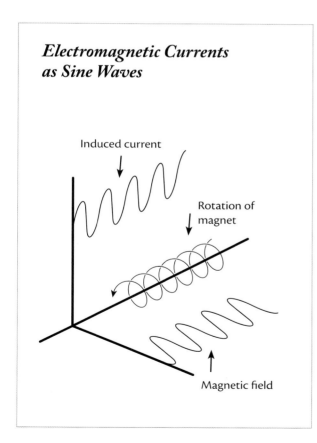

Electromagnetic Currents as Sine Waves

Induced current

Rotation of magnet

Magnetic field

Hippolyte Pixii: The First Commutator

The tendency of a rotary motion to create alternating current was demonstrated in practice when a French instrument maker named Hippolyte Pixii read of Faraday's discoveries and set about building a hand-cranked electrical generator in 1832. Pixii's design consisted of a crank attached to a horseshoe magnet that rotated so that the north and south poles of the magnet passed over a coil wrapped around an iron core. The free ends of the coil were connected to terminals that could detect the presence of an electrical charge. Pixii found that when the pole of the magnet would pass over the coil, it would induce a pulse of electricity. But the pulse changed direction each time a different pole of the magnet made its pass. What Pixii was observing was what we know today as alternating current.[11] Although he demonstrated his device at the French Academy of Sciences in 1832, like other inventors of his day, the twenty-four-year-old Pixii had no idea how to make his invention commercially successful. The few electrical machines in existence at the time ran off of batteries. Moreover, the most common use of electricity in the 1840s and 1850s was the telegraph, which sent signals by interrupting a direct current in a wire strung between a generator and a receiving station.[12] André-Marie Ampère (a fellow member of the Academy), however, suggested that Pixii fit his contraption with a device (now known as a commutator) that would turn the alternating current into more useful direct current. The commutator used a split-ring and copper wire brushes to reverse the direction of the current at the point that the rotating magnet started to return to a position parallel to the plane of the wire. In effect, the commutator turned the steady alternation of an induced current into a steady pulsation of current in a single direction. Instead of a sine wave, the commutator created a current flow that looked more like a series of humps. The strength of the current would pulsate, but never actually change direction.

Pixii exhibited the revised design in London in 1833.[13] Unfortunately, however, the young scientist died in relative obscurity only two years later and there is no evidence that he ever made a dime off of his generator.

Tesla's Eureka Moment (1881)

By Tesla's account, his design for a motor that did not require a commutator came to him in a bolt of inspiration: a truth hidden in the recesses of his ingenious mind, waiting to be revealed. Since his rebuke by Professor Poeschl in

Hippolyte Pixii's AC generator

1876, Tesla had claimed that he was constantly plagued by the challenge of creating a motor without a commutator. In other words, he was searching for another way to generate a steady rotary motion from an electrical current that constantly changed direction.

By his own admission, during his years in Graz, Tesla tried to work through the problem by contemplating various configurations of existing electrical generators. He would first start by picturing a DC generator (known as a "dynamo") in his head and running the system through various experiments using the power of his imagination. When that proved fruitless, Tesla claimed he would then imagine "an alternator and investigate the processes taking place in a similar manner."[14]

It is curious that Tesla would refer to the device he imagined as "an alternator," which is simply an electrical generator that converts mechanical energy into an alternating current. In truth, all dynamos start out as alternators, creating AC power. It is only because no one could figure out a way to make practical use of AC power that the standard design for a motor included a commutator, artificially converting AC to DC. An alternator, however, was little more than a dynamo, minus the commutator.

Around 1866, three engineers—Charles Wheatstone, Werner von Siemens, and Samuel Alfred Varley—all independently presented designs for the modern DC dynamo, though Varley was the only one to seek a patent. It was Siemens, however, who would begin commercial production of a DC generator he called a "dynamo" in the late 1860s. Then, Belgian inventor Zénobe-Théophile Gramme improved upon Siemens's design and, in 1871, began to manufacture the Gramme dynamo, the most popular design of that time. It was, in fact, one of Gramme's devices that Tesla witnessed in action during Professor Poeschl's classroom demonstrations.

The Gramme dynamo was capable of producing AC power, but the only popular use of AC at the time was in Yablochkov candles, a type of light bulb invented by Russian engineer Pavel Yablochkov and used to illuminate the Paris Exhibition of 1878. The term "alternator," in fact, did not even come into popular usage until around that time. So it is unclear how Tesla was running imaginary experiments on "alternators" as early as 1875–1876. Tesla biographer W. Bernard Carlson speculated that it was unlikely that, as a student in Austria, Tesla would have known about Yablochkov's Paris system. He concludes, therefore, that Tesla's inspirations must have come simply from scrutinizing how existing DC motors operated.[15]

Whether Tesla knew of the demonstration at the Paris Exhibition may never be known. The more likely explanation is that he did not actually start running thought experiments using an AC configuration until at least the mid-1880s, when alternators became the subject of several important scientific papers. The British electrical engineer James Edward Henry Gordon, for example, built two large-scale experimental alternators in 1882 while serving as assistant secretary for the British Association for the Advancement of Science (now the British Science Association). This was followed shortly by an even more powerful alternator designed and patented in 1886 by the eighteen-year-old British electrical prodigy Sebastian Ziani de Ferranti (though he would

The Paris Exhibition of 1878

The idea that Tesla may have known about Yablochkov's system (by word-of-mouth or through the popular press) might not be so far-fetched after all. Over thirteen million people attended the Paris Exhibition of 1878, and it brought more visitors to the city now known as the "City of Light" than any previous cultural event. Also, the Exhibition was held to celebrate the return of Paris to French control after Otto von Bismarck's siege during the 1870–1871 Franco-Prussian War. Just a few years earlier, the Prussians had struck a similar blow to the Austrians, who controlled the area of Croatia where young Tesla and his family (some of whom worked for the Austrian army) resided. Austria's waning influence surely was of interest to Tesla's father, who exhibited nationalist tendencies in several articles he wrote for Serbian newspapers and magazines in the 1850s. It is not entirely implausible, therefore, that Tesla and his family would have read with interest the widespread press coverage of the Exhibition, including the impressive lighting of the Avenue de l'Opera using Yablochkov's AC arc-lighting system.

Yablochkov's AC arc-lighting system illuminating the Paris Exhibition of 1878

later lose the patent rights when the design was found to have been anticipated by William Thompson—better known as "Lord Kelvin"—who actually had helped Ferranti construct the first operating model).

After his time in Graz (and a short sojourn in Gospić following his father's death), Tesla moved to Prague, ostensibly to enroll in Karl-Ferdinand University.[†] In Prague, Tesla claims to have made a "decided advance" in solving the problem of operating an electrical motor without the need of a commutator. According to Tesla, this advance consisted of "detaching the commutator from the machine and studying the phenomena in this new aspect..."[16] Tesla admitted, however, that detaching the commutator from a DC machine (essentially creating an alternator) did not solve the problem of "converting a steadily pulling force . . . into a rotary effort."[17] It merely left him where he started: with an alternator incapable of generating rotary motion. It is clear that, despite his uncanny ability to construct and run complex machinery in his mind, Tesla was no closer in 1880 to inventing a rotating magnetic field than were most of his contemporaries.

In January 1881, Tesla moved to Budapest, though it was not until the following year that he claims to have had the flash of insight that led to the invention of the polyphase AC motor. While recuperating from a deep depression, and experiencing strange and irritating bouts of extreme sense awareness (for which a doctor had

---

[†]   Though the Czech authorities can find no record of Tesla's having enrolled at Karl-Ferdinand or any other Prague school, biographer D. Mrkich claims that Tesla attended summer courses in mathematics, experimental physics, and at least one philosophy class on the noted empiricist David Hume. See D. Mrkich, *Nikola Tesla: The European Years* (Ottowa: Commoners Publishing, 2010), 92.

prescribed large doses of potassium), Tesla claims to have had his Eureka! moment.[18]

Anthony Szigeti—one of "a number of young men" in Budapest in whom Tesla claims to have "become interested"—insisted that the young inventor walk with him in Budapest's city park each evening.[19] Although Szigeti apparently was an accomplished mechanic, there is no evidence that he had any formal electrical training. Nevertheless, Tesla claims that during these walks they would discuss the young inventor's challenge in finding an improved electrical motor design.

During one notable walk, as the two gazed upon the setting sun, Tesla began to recite from memory verses from Goethe's *Faust*:

> *The glow retreats, done is the day of toil;*
> *It yonder hastes, new fields of life exploring;*
> *Ah, that no wing can lift me from the soil*
>
> *Upon its tracks to follow, follow soaring!*
> *A glorious dream! though now the glories*
>   *fade.*
> *Alas! the wings that lift the mind no aid*
> *Of wings to lift the body can bequeath me.*

Tesla does not clarify how these words related to the challenge that had been plaguing him for years. Nevertheless, upon uttering them, he writes that the idea of the polyphase electric motor came to him "like a flash of lightning."[20] Taking a stick from the ground, he claims to have drawn in the sand the diagrams below that would not appear publicly until six years later during an address Tesla gave before the American Institute of Electrical Engineers.

The Polyphase Design

Today, the fundamental concept of the polyphase motor seems so obvious that it is hard to understand why it was so revolutionary. It all begins with the basic law of magnetism: opposite poles attract and like poles repel. So if we imagine an electromagnet suspended so that it can rotate between the poles of a stationary magnet, it is easy to understand how the attracting and repelling forces can create rotational motion. The north end of the electromagnet is repelled from the north end of the stationary magnet and attracted to the south end. Similarly, the south end of the electromagnet is repelled from the south end of the stationary magnet and attracted to the north end. The electromagnet would spin until its poles aligned with the opposite poles of the stationary magnet.

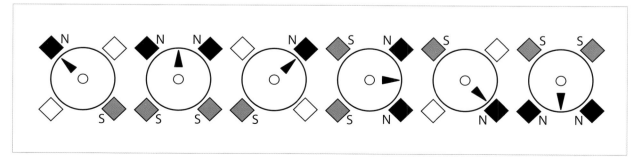

Polyphased electromagnet generates rotational motion

Under normal circumstances, motion would stop once the opposite poles aligned, at which point the electromagnet will have only made a half turn. Suppose that an alternating current allowed the poles of the electromagnet to reverse at predictable times. One could imagine the electromagnet spinning so that the opposite poles remained aligned. However, because an induced alternating current follows a sine wave, the strength of the current diminishes (and then reverses) as the poles align. The result is that the electromagnet is just as likely to reverse direction. Instead of a steady rotary motion, the "in phase" alternating current would cause the electromagnet to flip one way then the other.

But imagine if, instead of an electromagnet with two poles, the armature of a motor consisted of two magnets positioned at right angles so that their poles were ninety degrees from each other (see figure on page 95). Now, imagine that each of these electromagnets were charged by separate generators, each producing alternating currents timed just right so that the strength of the current charging a pole would pick up just as the strength of the preceding pole was starting to diminish. (In other words, the alternating currents charging each pole were slightly "out of phase" with each other.) The result is that the electromagnet would continue to spin toward the oppositely charged pole rather than in the reverse direction. Instead of flipping the electromagnet, the "polyphase" motor would propel it forward, in a steady rotation.

According to Tesla, these concepts were revealed to him so clearly that Szigeti, who had no formal electrical training, "understood them perfectly" as Tesla excitedly explained them. The epiphany had such a profound impact on Tesla at the time that he claims he could not "begin to describe" his emotions later. Still, somehow he found the words: "Pygmalion seeing his statue come to life could not have been more deeply moved. A thousand secrets of nature which I might have stumbled upon accidentally I would have given for that one which I wrestled from her against all odds, and at the peril of my existence."[21]

The emotional power of this revelation stands in stark contrast to the account Tesla provided during a patent dispute in 1903. Under oath, Tesla mentioned neither the stroll with Szigeti nor the remarkable insight that he claims hit him "like a flash of lightning."[22]

Szigeti apparently had no recollection of the event either. In an 1889 deposition to the State of New York, Szigeti never mentions Tesla's burst of brilliance in Budapest. Instead, he recalled that the inventor spoke of his motor design in Paris while the two were employed at Edison's French subsidiary in 1882.[23] Though Szigeti never mentioned Tesla describing a rotating magnetic field powered by alternating currents of differing phases, he did recall that Tesla was "much excited over ideas which he then [1882] had of operating motors . . . [and] dispensing with the commutator."[24]

The historical record appears to support Szigeti's account. Though Tesla's 1919 autobiography recounts the Budapest park scene, during his 1903 patent testimony, Tesla claimed to have drawn his diagrams in the dirt when explaining his idea to Edison coworkers in Paris sometime around 1883.[25] Regardless of when Tesla first conceived of the polyphase AC motor, the inventor insists that, within two months of this epiphany, he had conceived not only of all the designs for the motor, but also all of the modifications of the AC power distribution system for which he would become famous.[26]

Of Patents and Perjury

By the time Tesla received a patent on his motor design, he had already partnered with George Westinghouse to start manufacturing the device. But the two were by no means the first to market. Charles Brown and Mikhail Dolivo-Dobrovolsky, who had demonstrated long-distance transmission of AC power at the Frankfurt Electrotechnical Exhibition in 1891, were selling polyphase motors in Switzerland and Germany. In America, they were available from Elihu Thomson's company, Thomson-Houston, and from William Stanley, who had left Westinghouse's employ to start his own

Elihu Thomson

company.[27] So similar in concept were all of these designs that they spurred several court cases challenging Tesla's claim as the sole inventor of the polyphase AC motor. Though the U.S. courts would eventually rule in Tesla's favor, close examination of the historical record reveals that Tesla's greatest invention was more the brainchild of many fathers than a spark of brilliance by a lone inventor.

Westinghouse v. New England Granite (1900)

Despite any reservations Westinghouse may have had about the originality of Tesla's invention, he knew that he must start enforcing Tesla's patents in earnest, both to stave off the growing number of competitors and to recoup royalty payments on polyphase motors that had already been sold.[28] In 1889, Westinghouse turned his sights on New England Granite, a company that had been using Stanley's motors for cutting quarries in Connecticut and Massachusetts. Westinghouse sued the company in the federal district court in Connecticut, charging that it was infringing upon Tesla's patent by employing the inventor's motor design without paying the required royalties.

To defend itself against Westinghouse's patent infringement claims, New England Granite needed to defeat the validity of Tesla's patent. Essentially, it had to prove that the inventor neither discovered the idea of a rotating magnetic field powered by out-of-phase currents, nor was the first to suggest that it could be used to power an electric motor. The company relied principally upon two inventors—Walter Baily and Marcel Deprez—who had both presented scientific papers outlining the principles behind Tesla's polyphase motor years before the inventor's alleged epiphany in Budapest (and certainly well before he filed his patent application).

Walter Baily

In 1879, a British physicist named Walter Baily presented a paper before the Physical Society of London. Entitled "A Mode of Producing Arago's Rotation," the paper first described the experiment in which François Arago was able to make a copper disk rotate by spinning a magnetized needle suspended above it.[29] By the time of Baily's lecture, most scientists understood that the effect (known as Arago's rotation) was due to the spin of the magnetic field in which the disk was suspended. At the time, the only practical way to rotate the magnetic field was to mechanically rotate the magnet producing it.

But Baily thought that there might be another way to rotate the magnetic field that required no moving parts and no mechanical inputs. He surmised that one could alternate the strength of the poles of a stationary electromagnet fixed underneath the suspended disk. When, for instance, the strength of the positive pole increased, Baily believed it would attract the portions of the disk farthest from the pole. Likewise, as the strength of the negative pole diminished, it would exert less of an attraction on the portions of the disk closest to it. Thus, one could cause the disk to spin continuously in the same direction. Although genuinely unaware of it at the time, Baily had stumbled upon how to generate mechanical energy by the intermittent shifting of magnetic polarity, a key characteristic of Tesla's system for producing and transmitting electrical power.

The court, however, was unpersuaded that Baily's discovery undermined Tesla's claim to a patent on his AC motor design. In upholding Westinghouse's infringement claim, Judge William Kneeland Townsend noted in his 1900 decision that Baily's configuration included a commutator, while Tesla's motor did away with it altogether.[30] For Townsend, this distinction proved essential. Rather than focus his legal analysis on the single critical element that made Tesla's design novel—using alternating induction currents to achieve fluid, rotary motion—Townsend instead focused on the relatively nonessential difference in how the two designs harnessed the current they used.

Marcel Deprez

Though he never completed college, Marcel Deprez had a natural penchant for electrical engineering. Between 1876 and 1886, the Frenchman conducted a series of experiments trying to transmit power over long

Marcel Deprez

distances. Like his contemporaries, Deprez began with direct current (DC), attempting (unsuccessfully) to demonstrate a functioning long-distance DC distribution system at the Paris International Electricity Exposition in 1881. A year later, using higher voltages and multiple generation nodes, he was able to transmit 1.5 kilowatts of DC power from Miesbach, Germany, to Munich, some fifty-six kilometers (almost thirty-five miles) away.

In a series of reports presented to the French Academy of Sciences between 1880 and 1884, Deprez demonstrated mathematically what Tesla would produce mechanically: that a rotating magnetic field could be used to generate an alternating current. In one report, presented to the Academy in 1881, Deprez describes a device of his own invention (named the "annular comparer") that utilized commutator brushes to produce opposing polarities in copper windings surrounding an iron ring at right-angles to each other.[31]

Deprez noted that a magnet revolving within the ring would produce a shifting magnetic field, generating alternating currents with a corresponding phase difference of ninety degrees.[32] Although he speculated that such a configuration might be used to generate a small amount of power, his primary interest was in using the device as a new type of electric compass for maritime navigation.

Again, Townsend was unpersuaded. Even though Deprez had suggested, in theory, that his device could generate small amounts of power, the fact that his invention had never been *used* to produce energy was proof enough that Tesla's motor, "required invention to select . . . that particular kind of current which was necessary for the production of the best results, and to adapt the mechanism . . . to its practical application."[33] Here the critical question for Townsend was not who had originated the idea of an electric motor powered by alternating currents of multiple phase, but who could demonstrate its practicality. Admitting that Tesla was not the first to discover the rotary magnetic field, or even the first to suggest it could be used to generate mechanical motion, Townsend nevertheless ruled that "by a new combination and arrangement of *known* elements, [Tesla] produced a new and beneficial result never attained before."[34]

The Split-Phase Design

Despite Tesla's enthusiasm for the novelty of his motor design, his financial backers worried that it would prove impractical and expensive. Tesla's initial design relied on energizing each magnetic pole independently. Each phase of the motor, in other words, operated on its own circuit directly connected to a separate AC generator. Therefore, the system required between four and six different (and expensive) copper wires connecting two (or more) generators to the motor.[35] Alfred Brown, Tesla's chief funder at the time, was concerned that customers would stick with the cheaper DC motor despite the problems encountered trying to transmit DC over great distances. For all practical purposes, the Tesla Electric Company abandoned any work on manufacturing a polyphase motor until Tesla could devise a way of running it using less copper.

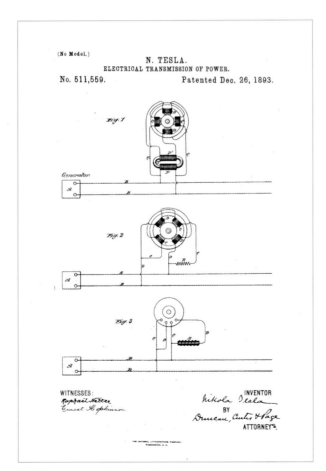

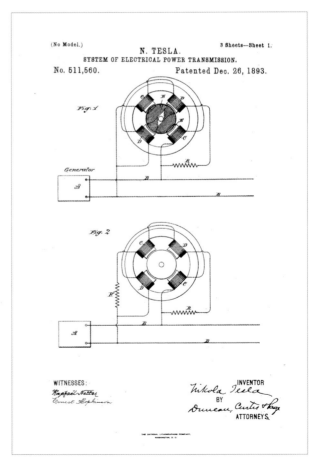

Tesla's patents on the split-phase design

Eventually, Tesla found a solution, though it is unclear how long it took him. In December of 1888, Tesla filed patent applications on a modification of his polyphase motor that would allow it to run on one generator.[36] Rather than having each pole independently energized by a separate motor (and slightly out of phase), this "split-phase" design used the alternating current from a single generator. By increasing or decreasing the number of times the wire is wound around the armature, it is possible to create the phase difference that was the foundation of Tesla's original design.

Tesla claims to have built several types of these "split-phase" motors in his Liberty Street lab in the fall of 1887. However, he said nothing about them until he visited his patent lawyer, James Page, in April of 1888. While drafting applications on Tesla's polyphase system, Page casually wondered aloud if Tesla's motor could operate with only two wires. Tesla responded that it could and proceeded to explain the split-phase design in detail.[37] According to sworn testimony taken several years later, Tesla had kept the design a secret because he worried the lawyer would think there was no great discovery in running a motor on a single circuit (like any common DC motor). Consequently, he feared if Page knew of the single-generator design, the lawyer would not appreciate its novelty and, hence, draft a weaker patent application on the original polyphase system.[38]

In his 2013 biography, Carlson speculated that Tesla may have viewed the split-phase modification as an inelegant intrusion upon the "ideal symmetry in his polyphase system" and that, "like other inventors, Tesla tended to have a blind spot about the commercial implications of his work."[39] The tendency to explain Tesla's strange behavior as a kind of odd quirk endemic to inventors does a disservice to the profession and rewrites history to comport with the myth of the lone inventor, whose bizarre quirks and foibles (no matter how irrational) set him apart from the less ingenious.

Rather than see Tesla's explanation for what it is—either suspicious or preposterous—many historians concoct, without a shred of primary evidence, an excuse that defies the obvious. If Tesla were truly concerned that Page's knowledge of the design would prompt the lawyer to write a weaker patent application for the polyphase motor, why would he disclose it so cavalierly when Page inquired? Of course, Tesla's explanation also requires that one believe that he drastically miscalculated Page's understanding of electrical innovations. Rather than dismiss the inventor's single-generator system as inconsequential, as Tesla feared, Page understood immediately its utility and quickly began the process of drafting patent applications on the split-phase design for Tesla to review.[40]

Westinghouse v. Dayton Fan & Motor Company (1901)

As Judge Townsend was in New York rendering his opinion, a federal circuit judge in southern Ohio, Judge Albert Thompson (who had served as a lieutenant in the Union Army during the Civil War), was considering a nearly identical suit that Westinghouse had brought against the Dayton Fan & Motor Company. In 1889, the company had started manufacturing an AC polyphase motor similar to Tesla's split-phase design.‡ Consequently, this 1901 case focused more on the inventor's claim to this modification than on Tesla's patent on the original motor design.

In its defense, Dayton Fan & Motor argued that achieving a phase-difference by lagging the current at the poles was no innovation at all. Any skilled electrician would understand that by changing the size or form of the wires charging the magnetic poles, one could adjust resistance to achieve a difference in phase between the poles without the need for separate generators.[41]

Like Townsend, Judge Thompson was less concerned with whether Tesla's split-phase design was commonly understood in theory and more concerned about whether anyone demonstrated the knowledge in practice prior to Tesla filing patent applications. In upholding Westinghouse's infringement claim, Thompson noted that, although "it is difficult to determine where skill ends and invention begins . . . neither [Dayton Fan & Motor] nor any one [sic] else" had constructed a split-phase system "until the way was pointed out by Tesla" in his 1888 patent filing.[42] Thus, rather than determine whether Tesla's modification was anticipated by the state of the art in 1888, Thompson abandoned much of the purpose of patent law and reduced innovation to the process of winning a race to the patent office. Since Tesla was the first to apply for a patent on the split-phase design, the court would regard him as the first to have thought of it (and so too would many historians).

---

‡   Incidentally, Dayton Fan & Motor Company changed its name to Day-Fan Radio in the early 1920s when it started producing radio parts. The company was bought by General Motors (GM) sometime in late 1929 or early 1930 and started making radios under the Delco label. However, an antitrust action brought by the Department of Justice forced GM to liquidate its radio holdings (including Delco) in 1939.

Dayton Fan & Motor Company quickly appealed Thompson's decision to the Sixth Circuit Court of Appeals. Judge Henry Franklin Severens wrote the opinion of a three-judge panel unanimously holding that the company had infringed both Tesla's patent on the polyphase motor and his patent on the split-phase design.[43] Interestingly, while affirming Judge Townsend's opinion in the lower court, Severens noted that Tesla's modification "greatly advanced his original invention for practical purposes."[44] Yet, he noted that his ruling rested, in part, on the legal rule that a second patent should be sustained if it is "fairly necessary" to the practical use of the first patent.[45] Practicality, in other words, played a contradictory role in the outcome of these two cases. In upholding Tesla's claim as the inventor of the polyphase motor, the *New England Granite* court dismissed Baily and Deprez's prior inventions as both theoretical and impractical. Yet, in upholding Tesla's claim as the inventor of the split-phase design, the *Dayton Fan & Motor* court acknowledged that Tesla's original motor was impractical, even suggesting that the split-phase modification was "fairly necessary" for the original motor to be of any use.

Westinghouse v. Catskill Illuminating & Power (1903)

Among the first companies targeted by Westinghouse when he launched his patent enforcement campaign in the late 1890s was Catskill Illuminating & Power Company, one of several upstate utilities eventually consolidated into the Upper Valley Electric & Railroad Company. In 1899, Westinghouse filed a patent infringement suit against the company in the federal circuit court for the Southern

District of New York.[46] The industrialist claimed that Catskill Illuminating had been utilizing Tesla's polyphase system as well as the inventor's split-phase modification. As in the New England Granite and Dayton Fan & Motor cases, the court in February 1903 reaffirmed Tesla's patents and held that Catskill Illuminating had infringed on both. The company decided to appeal the decision, at least in terms of the split-phase modification.

In the meantime, Theodore Roosevelt (who had risen to the presidency after the assassination of President William McKinley) promoted Townsend, author of the New England Granite decision, to the United States Court of Appeals for the Second Circuit, the court charged with hearing Catskill Illuminating's appeal. Thus, Judge Townsend was provided the unique opportunity to revisit Tesla's inventiveness.

Galileo Ferraris

The crux of Catskill Illuminating's appeal centered around a paper written by Italian physicist and electrical engineer Galileo Ferraris, who spent much of 1885 investigating the efficiency of the AC power system that the French inventor Lucien Gaulard and the British engineer John Dixon Gibbs had demonstrated at the Electrical Exposition of Turin, Italy, in 1884. Born nearly ten years before Tesla, Ferraris was one of four children of a successful pharmacist from a town just outside of Turin (then located in Sardinia, one of Italy's predecessor states). Ferraris attended the University and Engineering School in Turin, receiving a degree in mathematics and civil engineering in 1869 (about the same time that Tesla was impressing the townsfolk of Gospić with his heroic repair of their

fire pump). He went on to complete his master's degree and become a professor of physics at the Regio Museo Industrial in Turin.

In 1881, Ferraris was appointed Italy's representative on the awards jury of the first International Electricity Exposition in Paris (the same exposition that showcased Yablochkov's AC lighting system and Marcel Deprez's long-distance transmission of DC power).[47] The exposition was designed to showcase the many advances in electricity since electrical arc lighting had first been used to illuminate the Paris World's Fair in 1878. Upon returning from the Exposition, Ferraris founded a School of Electrotechnology (along with an extensive laboratory) at the Museo.[48] Two years later, he hosted an International Electrical Exposition of his own in Turin during which AC power was successfully

Galileo Ferraris

transmitted over forty kilometers, from Lanzo Torinese to Turin, using the transformer jointly designed by Gaulard and Gibbs.§

Like Faraday and Baily, Ferraris was interested in electrical motors that would convert electricity to mechanical energy. No doubt aware of Baily's article, in 1885 Ferraris first conceived of using two out-of-phase electrical currents to produce a rotating magnetic field. For three years, he worked on the design of a motor that used electromagnets powered by alternating currents—each ninety degrees out of phase—to spin a rotor without any additional moving parts. Moreover, Ferraris's design only required one generator to produce the multiple, out-of-phase currents needed to create the rotating magnetic field. On April 22, 1888, Ferraris presented a paper outlining his design to the Royal Academy of Sciences in Turin. It was quickly translated into English and published in the journal *Industries* later that same year.[49] By May, however, Tesla was granted a patent on his polyphase system (though he would not file an application for the split-phase modification until December 8, 1888).

Interested Witnesses

Upon reviewing Ferraris's paper, Judge Townsend determined that it fully described the split-phase system for which Tesla was granted a patent. Writing for a three-judge panel in 1903, Townsend reversed the lower court's decision

and held that there was insufficient evidence that Tesla had devised his split-phase modification prior to the original publication of Ferraris's paper on April 22, 1888.

In defending Tesla's patent, Westinghouse relied principally on three pieces of evidence: a photograph claimed to be of a split-phase motor in Tesla's Liberty Street laboratory in 1887; the testimony of Tesla's chief funder, Alfred Brown; and the testimony of James Page, who had filed Tesla's patent. (Tesla, presumably busy experimenting with wireless transmission in Colorado Springs, did not testify.) Townsend methodically reviewed and rejected each piece of evidence.

The photograph was the only piece of physical evidence showing that Tesla had constructed a motor with a split-phase modification prior to filing his patent in December 1888. According to Brown's testimony, all of Tesla's prototypes had been moved to the inventor's Fifth Avenue laboratory, which had burned to the ground (along with its contents) in 1895. For Townsend, however, the photograph was practically useless. The motor it depicted looked no different than Tesla's original polyphase motors, and nothing in the photograph indicated that it was adapted to run on a single generator.

Although Brown had sold his interest in the Tesla Electric Company by 1903, in 1900 (while he still retained a financial interest in the outcome of the case) he testified that Tesla not only had explained the split-phase design to him in detail, but had shown him several split-phase motors in operation at the Liberty Street laboratory before the inventor had moved his operation to South Fifth Avenue in the summer of 1892. Townsend noted, however, that Brown could neither identify the date of Tesla's disclosure, nor

---

§   Gaulard and Gibbs actually had first demonstrated their transformer design in London in 1881. Although there is disagreement as to whether George Westinghouse was present or had merely read about the invention in the engineering literature, the American industrialist nonetheless was so impressed with the Gaulard-Gibbs design that he bought it from them, had improvements made by William Stanley (who had not yet left Westinghouse to start his own company), and patented the resulting "Stanley Transformer" in 1886 (see Chapter Two).

could he provide anything more than a "vague general description" of the apparatus. What explanations Brown could recall of Tesla's alleged disclosure, moreover, appeared contradictory.

For its part, Catskill Illuminating presented evidence that Tesla had spent nearly two weeks in 1888 at the Mather Electric Company, a Connecticut-based power company serving Hartford and Manchester, trying to convince them to manufacture his polyphase motor. Catskill introduced evidence that technicians at Mather objected to Tesla's design on the basis that it required an extra generator and several more copper wires than the standard DC setup. Curiously, Tesla never mentioned to the engineers at Mather that he had already designed a version of the motor that would not require these added expenses.

Carlson speculated that Tesla was operating under instructions from Brown, who did not want the inventor to reveal everything he had accomplished.[50] But this explanation is problematic, given that Carlson admitted that the Mather Electric Company was Tesla's first choice for partnering on the manufacture of his motor. Carslon even noted that Tesla felt William Anthony (a noted electrical engineering professor who left Cornell University in 1887 to become Mather's chief engineer) was the best person to help improve the motor designs.[51] In March of 1888, Brown had even dispatched Tesla to Mather's headquarters in Manchester solely so Anthony could examine Tesla's polyphase motors. At the time, Anthony apparently registered little concern about the extra wiring because he assumed Mather would be marketing Tesla's polyphase motors for special industrial purposes, where

the cost would be less of an issue.[52] However, once Mather indicated that the polyphase motor would be too expensive to market to the general public, it would have made little sense for Tesla to continue concealing his split-phase design. Carlson does not explain why Tesla did not reveal his improved version, instead choosing inexplicably to sacrifice a partnership with the company he most favored.

Page claimed that he, too, had seen Tesla's prototypes in 1887 and 1888. However, he was forced to admit under questioning that Tesla had "never intimated in any way" to him directly that he had invented a design that did not require the use of independent generators. In fact, Page's entire recollection of Tesla's disclosure rested on an entry in the patent lawyer's diary (which he could not produce) in which he recalled having recorded "services for this particular matter" sometime between April 8 and April 18, 1888.[53] Townsend summarily dismissed this testimony, noting not only that there was nothing to corroborate Page's recollection (not even the diary), but that the lawyer had filed patents on Tesla's other inventions in both May and October of 1888. The fact that Page did not file Tesla's application for the split-phase design until December 8, Judge Townsend observed, suggested that the "particular services" Page rendered more likely referred to these other inventions and not to the split-phase design.[54]

Tesla's Testimony: Westinghouse v. Mutual Life Insurance (1904)

Westinghouse's legal defeat was a bombshell. Following the court's decision in the Catskill Illuminating case, other Westinghouse targets began relying on the Ferraris article for their

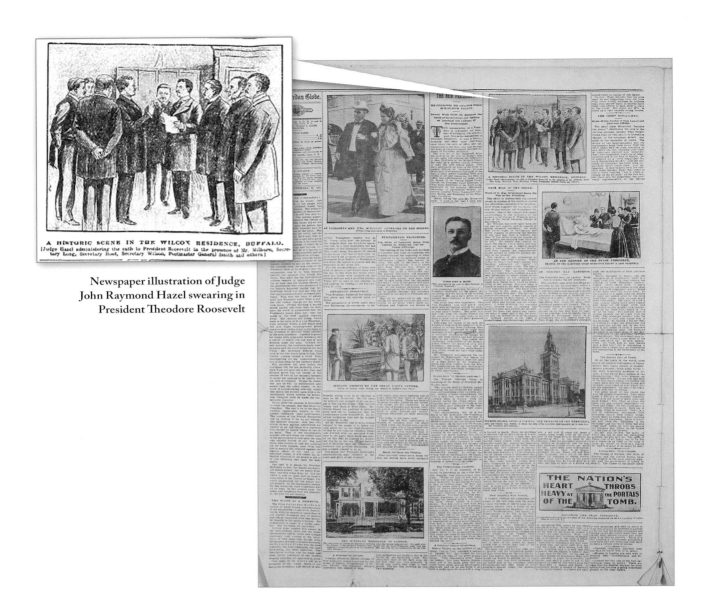

Newspaper illustration of Judge
John Raymond Hazel swearing in
President Theodore Roosevelt

defense, including the Stanley Instrument Company, which was churning out motors almost identical to Tesla's.[55]

Once the intrepid inventor returned from Colorado, however, everything would change. Westinghouse brought a suit against the Mutual Life Insurance Company in the federal court for the Western District of New York, a court within the jurisdiction of the Second Circuit Court of Appeals. Judge John Raymond Hazel, a McKinley appointee with close ties to

Roosevelt, reversed the upper court's decision and ruled that Tesla's patent on the split-phase design was valid after all.[56]

Under the legal doctrine of *res judicata*, Westinghouse would normally have been barred from bringing a suit based on claims that the court had already reviewed and sustained on appeal. However, in a lengthy ruling in favor of Westinghouse, Judge Hazel was quick to note that this doctrine only applied "provided no new evidence upon the subject is shown."[57] The soundness of his legal reasoning aside, Judge Hazel's decision to reopen the case allowed his court to reconsider (and ultimately reject) Townsend's analysis.

In reaching a conclusion entirely different than Judge Townsend's, Hazel relied principally on two pieces of new evidence: the mysterious appearance of a prototype split-phase motor (that Tesla allegedly constructed prior to the publication of Ferraris's paper) and Tesla's own testimony that he had built the motor sometime in February 1887. Given this new evidence, where Judge Townsend saw unsupported and inconsistent witness statements, Judge Hazel now saw corroboration of Tesla's claim. According to Tesla, the prototype suddenly produced for the Mutual Life case survived because it had been sent to the Patent Office in Washington, D.C., sometime prior to the 1895 fire that destroyed the inventor's Fifth Avenue laboratory.[58] For Hazel, Tesla's testimony, "emphatically and unequivocally narrated" by the inventor¶ and further supported by Brown, was sufficient to conclude that Tesla had thought of the split-phase design before Ferraris had presented it.[59]

Judge Hazel apparently did not find it odd that Page—who was responsible for filing Tesla's patents—made no mention in his previous testimony of the prototype motor he presumably had delivered to the Patent Office himself. For

his part, Tesla offered no documentary evidence that the subject apparatus had been sitting on a shelf somewhere in Washington while more than a dozen court cases spread over nearly a decade considered the matter.

Tesla's claim was supported only by the same testimony from Brown that Townsend had found vague and contradictory. Noting that Brown's financial interest in the outcome of the case "undoubtedly tends to detract" from the force of his testimony, Hazel nonetheless concluded that there was "no sufficient reason" to disregard it.[60]

As for Tesla's failure to mention the split-phase design to officials at the Mather Electric Company, Tesla testified that Brown and Page ("both financially associated with him") had instructed him to keep the invention a secret.[61] Hazel accepted this explanation as "significant fact" without questioning why, during previous testimony, neither Brown nor Page had mentioned that they had given Tesla this instruction. Such testimony would have been invaluable in lawsuits seeking to establish Tesla's priority because it would have provided some plausible explanation for why the only early witnesses to Tesla's split-phase system were individuals with a financial interest in the design.

Mutual Life introduced new evidence of its own, including testimony from William Stanley that he had visited Tesla's laboratory between May 15 and June 15, 1888, and neither saw the prototype motor, nor recalled Tesla mentioning anything about an improved design.[62] Hazel dismissed this testimony on the basis of two (new) facts he found compelling. First, he referenced a June 24, 1888, letter from Tesla to Westinghouse (written some two months after Ferraris's publication) explaining how the split-phase

¶   Since at least 1997, federal courts have repeatedly held that an inventor's own testimony is insufficient to establish the date of invention and must be independently corroborated. See *Fina Oil* and *Chemical Co. v. Ewen*, 123 F.3d 1466 (Fed. Cir. 1997). This requirement incidentally, stems from a historic distrust of an inventor's uncorroborated oral testimony. See *Woodland Trust v. Flowertree Nursery, Inc.*, 148 F.3d 1368 (Fed. Cir. 1998). It is doubtful, therefore, that Tesla's testimony—no matter how emphatically narrated—would constitute sufficient evidence of prior invention under modern patent law.

modification worked.[63] Although Brown and Page had both instructed the inventor to keep the design secret, apparently this admonition did not apply to Westinghouse (who, incidentally, was not yet financially invested in Tesla at the time the inventor claims to have written the letter).[64]

Curiously, Carlson documented a visit Stanley (who was working for Westinghouse at the time) made to Tesla's laboratory on June 23, 1888. According to Carlson, Tesla's other business partner, Charles Peck, suspected that Stanley was working on an AC motor design of his own and instructed Tesla—counter to all of the previous concerns about secrecy—to show him the split-phase motor in an attempt to "counter any claims by Stanley that he has invented a motor better than Tesla's."[65]

Although Carlson implied that the decision to suddenly reveal the split-phase design was part of an elaborate plan to put Tesla's business partners in a better negotiating position with Westinghouse, that explanation makes little sense given Tesla's willingness to conceal the design during negotiations with Mather. It also flies in the face of Peck's own concern that Stanley was busy designing his own AC motor, especially as it was nearly six months before Tesla would apply for a patent on his split-phase design. Indeed, Seifer noted that, when Stanley reported back to Westinghouse, he claimed that Tesla's motor was nothing new. Instead of extolling Tesla, Stanley pointed his employer to his own notebook entries from September 1882 in which the engineer had outlined an AC power distribution system nearly identical to Tesla's design.[66] Given Stanley's report, either he was referring to having previously envisioned

the split-phase design or he was referring to the polyphase system more generally (implying Tesla had not yet revealed to Stanley the modification that would make his motor commercially viable).

If Peck worried that Stanley would steal Tesla's split-phase innovation before the inventor was able to file for a patent, it stretches credulity to think that he would instruct Tesla to reveal the design in a misguided—and, according to Seifer, unsuccessful—attempt to intimidate the engineer into abandoning his own motor and recommending Westinghouse buy Tesla's instead. In any event, it appears that Westinghouse was already aware of Ferraris's paper and was worried enough about Tesla's patent claims that, sometime during his negotiations with Brown and Peck in 1888, he sent an agent to Italy to purchase options on Ferraris's design.[67]

Second, Hazel determined rather counterintuitively that the inability to locate Anthony Szigeti—who was Tesla's assistant at the time the inventor claimed to have devised the split-phase design—tended "to explain [the assistant's] failure to corroborate" Tesla's claims."[68] Evidence of absence for Hazel, in other words, was not absence of evidence. Rather, the assumption was, counter to all legal doctrine, that Tesla's claims were true until proven otherwise. The court's inability to locate Szigeti, however, is made even more mysterious by conflicting historical claims about the mechanic's whereabouts. In his 2013 account, Carlson wrote that Szigeti left Tesla in 1891 to pursue his own inventions, perhaps in South America.[69] But, in Seifer's exhaustive 1998 biography, the historian claimed Szigeti died sometime over the summer of 1890. He cited an August 18, 1890, letter from Tesla to his family in which the inventor noted Szigeti's passing and how it made him feel "alienated" in the New World.[70] If (as is likely) the latter account is the more accurate, one wonders why no one mentioned that Szigeti was dead when the court sought his testimony.

The Patent Pool

When Tesla's patents on the split-phase motor expired on May 4, 1905, Westinghouse's old nemesis (and former employee), William Stanley, was among the most relieved. In the pages of *Electrical World and Engineer*, Stanley speculated that the courts never would have maintained the validity of Tesla's patent claims if not for the financial interest of their owners, a "Patent Pool Trust" including George Westinghouse, William Waldorf Astor, and by 1890, even Thomas Edison's company, General Electric.[71] By early 1891, in fact, Westinghouse was facing financial ruin. The failure of a major London brokerage house, Baring Brothers, created a panic among Westinghouse's creditors, many of whom called in their loans.[72] The Westinghouse Company was forced into receivership and Westinghouse spent several years trying to regain control of his own enterprise. Tesla, in an attempt to help Westinghouse survive through this credit crunch, relinquished his right to royalties on his polyphase and split-phase patents.[73] Tesla's sacrifice certainly helped, although it ultimately denied the inventor billions of dollars that might have avoided his own bankruptcy. However, Westinghouse's liquidation strategy included a fail-safe: joining forces with his rivals, including Edison's General Electric (which, by that time, was being backed by none other than J. P. Morgan) and Elihu Thomson's Thomson-Houston Company.[74]

By 1895, rumors were circulating that, despite the public rivalry between Westinghouse and Edison, the two had reached a tentative agreement to pool patents.[75] Although it apparently took several years for Westinghouse, Morgan, and Edison to settle the arrangement, the deal they agreed to allowed General Electric (which had merged with Thomson-Houston in 1892) to use Tesla's patents on the polyphase motor, while Westinghouse would be permitted to use Edison's patents on various trolley systems.[76] The patent pool arrangement gave General Electric 62.5 percent of the value of all the pooled patents, while Westinghouse would retain 37.5 percent. If either company exceeded sales of its share of a particular product, it paid royalties to the other in order to maintain the agreed-upon division.[**]

Once the deal was inked, the pool turned its attention toward buying out competitors like William Stanley's company (which was absorbed by General Electric in 1903) or forcing them out of business through aggressive enforcement of Tesla's patent on the polyphase system.[77] By 1898, Tesla's patents on the split-phase design were also included in the patent agreement, ensuring that, by 1903, the collective influence of the nation's most powerful financiers and industrialists would be brought to bear in upholding Tesla's claim to have invented the motor prior to Ferraris.[78]

[**] Although popular accounts pit Edison against Tesla (and Westinghouse), the patent pool agreement makes clear that it was Westinghouse who ultimately thwarted Tesla's commercial success.

Although he profited from the merger with General Electric, Stanley accused the owners of Tesla's patents of waging a systematic campaign to bring "the great weight of public opinion" upon the courts that would likely adjudicate the issue. According to Stanley, "their efforts to fasten attention upon the inventor and his work were ably supplemented by numerous, startling statements, promises, and disclosures" by Tesla himself. [79]

Although no one has uncovered actual evidence of chicanery, there were few degrees of separation between Tesla (and his financial backers) and the judge who would upend legal precedent to find the inventor's split-phase patent valid. Tesla's social circles included political elites and the financial interests they often served. Robert and Katherine Johnson, the closest to a family Tesla had in New York, had known Theodore Roosevelt since before he ran for the New York State Assembly in 1881. [80] Roosevelt and Stanford White, one of Tesla's best friends, socialized together as early members of the Automobile

Tesla with Westinghouse associates

Club of America.[81] Roosevelt's own sister, Corinne Robinson, was among Tesla's cadre of women confidantes.[82] As a delegate to the 1900 Republican National Convention, Judge Hazel was among those who nominated then-governor Theodore Roosevelt as vice president. He even presided over Roosevelt's swearing-in after McKinley's assassination in 1901. Though no evidence of untoward influence has ever surfaced, the implication was enough for Stanley to conclude that the courts, led by Roosevelt's hand-picked justice, "quite innocently swallowed the bait prepared for them" and rendered a verdict favorable to Tesla's financial backers, regardless of the merit (or lack thereof) of the evidence.[83]

"My progress was so rapid as to enable me to exhibit at my lecture in 1891 a coil giving sparks of five inches . . . Since my early announcement of the invention, it has come into universal use and wrought a revolution in many departments."

—NIKOLA TESLA, *MY INVENTIONS*, 1919

chapter five

Electrical Transformers and the Tesla Coil

In the winter of 2010, the Federal Emergency Management Agency (FEMA) and a handful of senior military officials met quietly in Boulder, Colorado, to run a simulation. FEMA wanted to know how the power grid would react if North America was struck by a particularly severe solar flare, similar to the largest on record (which struck Canada in 1859). The results were sobering. According to the best experts, the storm would induce massive current spikes across the Northern Hemisphere sparking cascading blackouts as far south as Atlanta, Georgia. The simulation revealed that a storm of the size known to hit the earth about once every hundred years could cripple the U.S. power grid, causing up to $2 trillion in economic damage, with power outages lasting up to six months or longer.

The key to the calamity was the large number (up to 350) of extra-high-voltage transformers—the kinds that help transmit power from, say,

Quebec to New York City—that would become saturated with solar-induced currents, causing overheating and permanent core damage. Most utilities lack back-up transformers and could not turn to spares to replace those burnt out by a particularly strong solar storm. To make matters worse, almost all transformers are manufactured overseas. Because of the high demand for replacement transformers across many countries, U.S. orders could be backed up for up to six months. Meanwhile, critical parts of the electricity grid would be rendered essentially useless.

The world's bulk power transmission systems are completely dependent on transformers. They are the critical components in long-distance transmission of AC power. Without them, it is nearly impossible to transform large currents into high enough voltages to overcome line losses. The government officials tasked with responding to this kind of crisis were shocked. They shouldn't have been.

OPPOSITE: Tesla in front of the spiral coil, 1896

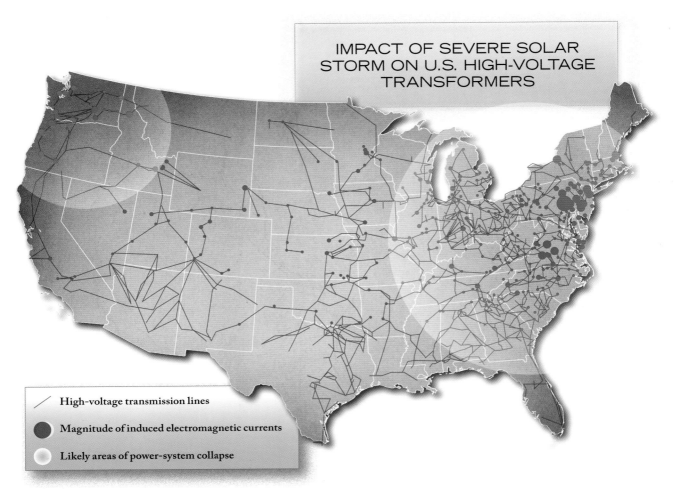

IMPACT OF SEVERE SOLAR
STORM ON U.S. HIGH-VOLTAGE
TRANSFORMERS

/ High-voltage transmission lines

● Magnitude of induced electromagnetic currents

○ Likely areas of power-system collapse

The alarming reality is that our entire system of transmission and distribution of alternating current hinges on a technology few people think about and even fewer understand.

Understanding Transformers

At its most basic, a transformer is simply a device that uses electromagnetic induction to transfer energy from one circuit to another. Recall that every electrical current generates a corresponding magnetic field, and every magnetic field is capable of inducing an electrical current. Michael Faraday discovered that an initial current generates a magnetic field that can, under certain conditions, induce a secondary current (see page 76).

Although Faraday was the first to articulate (and publish) the laws of electromagnetic induction, the process was first observed at an all-boys prep school nearly a year earlier. In 1829, Joseph Henry, a young professor of mathematics and science (then called "natural philosophy") at the Albany Academy, was experimenting with electromagnets, trying different configurations in an attempt to increase their power. Early electromagnets were of the most basic design, an electrified copper wire (now known as a "winding") wound around a soft iron core. Scientists had already experimented with increasing the strength of the magnet by increasing the number of turns of the wire around the core. But they quickly found that they reached a certain threshold when the turns got too close together. By insulating the wire, Henry discovered that he could prevent the electrical current from passing between the turns and, therefore, could generate a much stronger magnet.[1] The reason is fairly simple. Since the magnetic fields generated by each turn of the wire pass through the center of the coil (and into the iron core), more turns will generate a stronger magnetic field.[*]

Induction Coils

Inevitably, someone would discover what happens if you wrap two separate wires around a common magnetic core. That someone was a little-known Irish priest and scientist named Father Nicholas Joseph Callan. Where Henry wanted to design a stronger electromagnet,

---

[*] Keep in mind the "Right Hand Rule" (p. 76). Imagine that the fingers of your right hand represent turns of the wire. Your thumb will indicate the direction of the magnetic field lines passing through the iron core and also, incidentally, point toward the north pole of the resulting electromagnet. The general theory indicated here is that the more fingers you have curled around the core, the stronger the magnetic field generated in your thumb.

Father Nicholas Callan

Callan (after reading of Faraday's discovery of induction) wanted to generate stronger electromagnetic currents. In 1836, in a basement laboratory at Maynooth College in Ireland, Callan took a bar of soft iron and wrapped two separate lengths of copper wire around it, allowing the ends of each wire to hang independently. Then, he connected a battery to the starting end of only one of the wires. After some fiddling, Callan found that every time he broke the connection between the battery and the first wire, an electrical spark jumped between the ends of both wires.[2] He thought he had created a new kind of electromagnet, but what he actually designed was the first induction coil (or "inductor"), a primitive type of transformer.

In the 1840s and 1850s, several scientists experimented with Callan's design to generate pulses of high-voltage electricity from a relatively low-voltage supply of direct current (either from a battery or a dynamo, the DC power generators that were becoming popular around the same time). They found that, by increasing the number of turns of the second copper wire, the spark generated each time the power supply was interrupted had a much higher voltage than the electrical current supplied to the first wire.

What exactly was causing the voltage jump? When a direct current passes through the first wire (called the "primary winding"), it generates a corresponding magnetic field. Because the wires share a common magnetic core, any magnetic field generated in the primary winding will become coupled with the second wire (called the "secondary winding"). When the supply of direct current is interrupted, the magnetic field undergoes a rapid change as it collapses. As Faraday discovered, a changing magnetic field induces a

The Ruhmkorff Lamp

In 1870 Jules Verne published *Twenty Thousand Leagues Under the Sea*, a classic science-fiction tale about the adventures of Captain Nemo and his submarine, *Nautilus*, which was lit by a "Ruhmkorff Lamp." These early portable electric lamps were named for the German instrument maker Heinrich Daniel Ruhmkorff, who manufactured the first commercially available induction coils. The Ruhmkorff lamp consisted of a battery-powered induction coil that would generate a spark inside a vacuum tube filled with a rarified gas like neon, argon, or mercury vapor. A kind of crude neon light, these tubes would glow various colors corresponding to whatever gas was electrified by the induction coil. Although Verne popularized the Ruhmkorff lamps, they were actually first developed for miners by Alphonse Dumas, a French engineer who worked at the iron mines at Saint-Priest.

Ruhmkorff Lamp

secondary electrical current. Thus, each time the initial current is removed, a pulse of induced electricity is generated by the magnetic field shared by both wires. This pulse jumps across the gap of air between the ends of each wire, creating a spark as it does so.[3] Consequently, the first induction coils were known as "spark coils" and were classified by the length of the spark they were able to produce.[†]

Electromagnetic induction requires changes in the magnetic field generated by an initial current. A DC power supply, therefore, has to be interrupted repeatedly in order to generate a secondary electrical current. Consequently, early induction coils used a vibrating arm called an "interrupter" to rapidly break the connection between the power supply and the primary winding.

Turns and Voltage Ratios

Although some enterprising inventors were able to design interrupters that could break direct current connections many hundreds of times per second, alternating current could generate the required change in magnetic field—and at much higher frequencies—without the need for an interrupter. This characteristic proved to be extremely advantageous when transmitting AC power over long distances.

Recall that power is the product of current and voltage (p. 84). By transforming AC power into higher voltage (but lower current), we can reduce substantially the amount of line loss

---

†   While his invention was not as impressive as Tesla's five-inch coil, Heinrich Ruhmkorff patented an induction coil that routinely achieved sparks up to two inches in 1851, some forty years before Tesla. See Charles Grafton Page, *History of Induction: The American Claim to the Induction Coil and Its Electrostatic Developments* (Washington, D.C.: Intelligencer Printing House, 1867), 104–06.

due to resistance. An induction coil (or "transformer") uses the magnetic field shared by the primary and secondary winding as a way to convert the form of power. According to the right-hand rule, as current flows through the primary winding, it generates magnetic field lines that run perpendicular to the flow of current. As these field lines expand outward, they cut across each turn of the secondary winding. Just as an electromagnetic iron bar increases in strength with the number of turns of copper wire around it, so too does the number of turns in the secondary winding determine the strength of the voltage induced.

The ratio between the number of turns in the primary winding and the number of turns in the secondary winding will determine the total voltage generated in the secondary winding.[4] Assume that the magnetic field generated by an alternating current expands evenly through the core of an induction coil. Under these circumstances, magnetic field lines cut evenly across each turn of the secondary winding. As a result, the voltage induced in each turn is the same. The total voltage generated in the secondary winding, therefore, increases with the number of turns. More turns generate higher voltage; fewer turns generate lower voltage.

Consider the simple transformer on the left. The primary winding consists of ten turns around a common core. The secondary winding consists of two turns around the core. The ratio of the voltage (5:1) is equal to the ratio of the turns

Simple 5:1 Step-Down Transformer

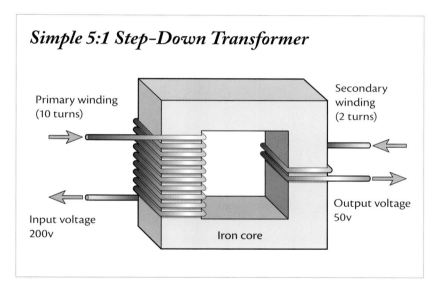

Primary winding
(10 turns)

Secondary winding
(2 turns)

Input voltage
200v

Output voltage
50v

Iron core

Components of a Simple Induction Coil

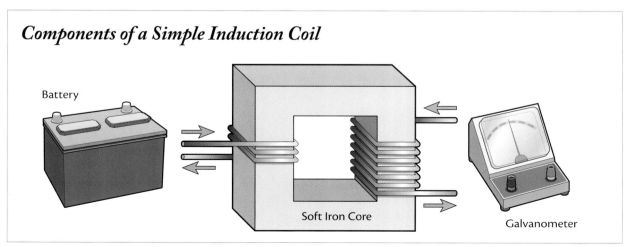

Battery

Soft Iron Core

Galvanometer

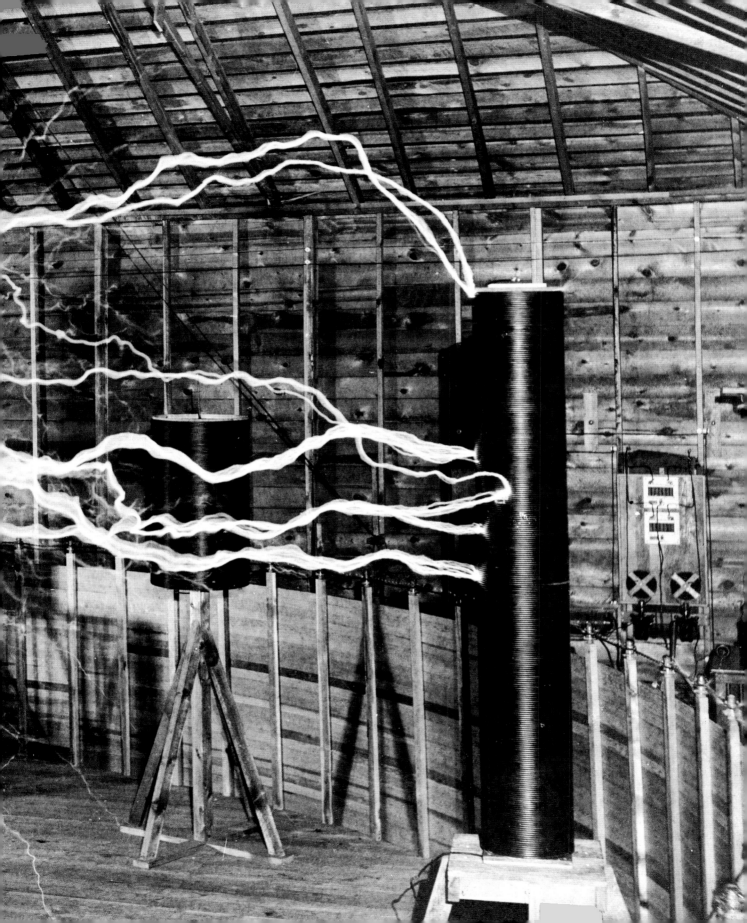

(10:2). Thus, the voltage of power introduced into the primary winding will induce a voltage one–fifth as strong in the secondary winding. Because the voltage induced in the secondary is less than the voltage introduced in the primary, this configuration is known as a "step-down" transformer. Similarly, a transformer in which the ratio of turns generates a higher voltage in the secondary winding is known as a "step-up" transformer.

Electromagnetic Radiation

In 1865, around the time a young Nikola Tesla was settling into his new home in Gospić, Croatia, a well-to-do Scottish physicist named James Clerk Maxwell published "A Dynamical Theory of the Electromagnetic Field." In the article, Maxwell made a revolutionary proposition: that both electric and magnetic fields, like light, travel through space at the same speed. Moreover, he proposed that all three essentially were various forms of the same phenomenon, different kinds of vibrations traveling through the ether at the speed of light, but at different frequencies.[5]

Maxwell derived his theory mathematically, inspired in part by his initial calculation that an electrical field propagates at the speed of light. The physicist suspected that this was more than coincidence.[6] By 1891, many of the theories Maxwell had derived mathematically were confirmed through experiments. Relying on Maxwell's equations, in 1888 the German physicist Heinrich Hertz engineered and tested equipment designed to transmit and receive electromagnetic pulses (known today as radio waves). After accounting for all other known forms of wireless energy, Hertz concluded that

PREVIOUS SPREAD: The long electrical arcs in this multiple exposure photo of Tesla sitting in his Colorado Springs laboratory were artificially generated for effect. **TOP:** James Clerk Maxwell **ABOVE:** Heinrich Hertz

his experiments definitely established that light, electrical fields, and magnetic fields were all forms of electromagnetic radiation, waves traveling through space carrying radiant energy at the speed of light.[7] In attempting to prove Maxwell's theory, Hertz unwittingly created the first radio transmitter and receiver, though he never patented the instruments as such. He theorized that electromagnetic waves might induce an electrical charge in a wire conductor whose length allowed it to oscillate at a frequency that resonated with the frequency of the electromagnetic wave. Consequently, the ingenious German created a simple device—a loop of wire containing a small spark gap—that would generate a spark discharge when in the vicinity of a resonant electromagnetic field.

Like his contemporaries, Tesla was fascinated by Maxwell's novel theory unifying light, electricity, and magnetism. He began thinking about electricity as vibration. Hoping to create lamps of unprecedented brightness, the inventor set about designing a device which could generate high voltage currents that alternated at frequencies approaching those of visible light (which he calculated to be around five hundred trillion vibrations per second).[8]

The Oscillating Transformer: Tesla's Coil

On May 26, 1890, Tesla filed the first of a series of patents that introduced a new type of transformer that he called an "oscillating transformer," but which would come to be known more famously as "the Tesla coil."[9] The device essentially was a step-up transformer with three major modifications: a large spark gap separating the primary and secondary windings,

Transformers and Climate Change

Just like power lines, the copper windings of a transformer give some resistance to the flow of current within them. Also like power lines, this resistance generates heat. In most cases, the heat simply dissipates with the cooler ambient air temperature. However, in some large transformers, resistance can generate so much heat that some sort of active cooling mechanism is necessary to prevent both the copper windings and the iron cores from being damaged.

Early transformer technicians found that mineral oil acted not only as an insulating fluid, but also as a fairly effective coolant. In fact, mineral oils specifically for use in electrical transformers were commercially available as early as 1899.[10] Around 1936, engineers started using a class of fluids known as polychlorinated biphenyls (PCBs), mostly because they were less flammable than mineral oil. Unfortunately, in the 1970s, the Environmental Protection Agency determined that PCBs were both carcinogenic and environmentally toxic. Their use in utility equipment was banned in 1977.

As PCBs were coming under scrutiny, the electricity industry turned to a chemical coolant that appeared to possess the flame-resistance of PCBs without their toxicity. Sulfur hexafluoride (SF_6) is chemically inert and is efficient at absorbing heat inside a transformer. Introduced in the 1960s, SF_6 became the coolant of choice after the ban on PCBs.[11] It turns out that SF_6, for all of its advantages, is the most potent greenhouse gas yet to be discovered, with a potential global warming effect 22,000 times that of carbon dioxide.[12] Given the danger that the chemical compound may escape transformer casings and contribute to global climate change, California has restricted the use of SF_6 and the European Parliament has called for its ban altogether.

Arc and Fluorescent Lighting

The first truly electric light was the carbon-arc light invented by Humphry Davy sometime between 1803 and 1809. Arc lighting refers to light created by the ionization of gas from a high-voltage discharge, for example, in a spark gap. In Davy's case, he used charcoal sticks connected to a 2,000 cell battery to create a discharge arc across a 4-inch spark gap.[13] By the 1870s, arc lighting replaced gas for lighting streets and public plazas.

In 1856, a German glassblower named Heinrich Geissler created a vacuum pump that allowed him to evacuate a tube of mercury gas to an extent not previously possible. When he passed an electric current through these "Geissler tubes," he could obtain a bright green glow from the walls of the tube at the end where an electrode was inserted. As vacuum tubes became more efficient, several scientists began experimenting to determine how various substances might be excited inside a Geissler tube.[14] Thomas Edison,

Geissler gas discharge tube

for example, used a coating of calcium tungstate to create a fluorescent bulb that he patented in 1907 (although it was never manufactured). This was the dawn of fluorescent lighting.

In the 1890s Tesla made similar experiments with fluorescent bulbs excited by high-frequency alternating current, though, like Edison, none was commercially successful.[15] However, at the 1893 World Columbia Exposition in Chicago, Tesla displayed a few of these devices at the Westinghouse exhibition space.[16] This event has led many to falsely credit Tesla as the inventor of fluorescent lamps.[17]

An American engineer and inventor, Daniel McFarlane Moore, developed a predecessor of the fluorescent lamp, called the Moore Tube, in 1896. The tube looked very similar to today's fluorescent bulbs, except that it was longer and used carbon dioxide and nitrogen as the excitable gas, creating an eerie pink glow. Moore achieved some success in the early 1900s selling the tubes to department stores in the New York City area.

In 1901, another American inventor, Peter Cooper Hewitt (son of New York Mayor Abram Hewitt), received a patent on the first mercury-vapor lamp, widely regarded as the forerunner of today's fluorescent lights.[18]

introduction of a capacitor between the power source and the primary winding, and synchronization of the discharge of the capacitor and the pulse of induced current at a resonant frequency. Together, these three improvements resulted in a remarkable device. But a close review of the historical record reveals that none of these innovations was novel. Although Tesla received patents on the various components, they were simply applications of scientific principles and

electrical configurations already discovered by a number of scientists with whom Tesla interacted over several years prior to finalizing his patent.

Large Spark Gap

Tesla's coil worked in two stages. First, Tesla reconfigured a standard step-up transformer with a large air gap between the primary and secondary windings. In a conventional transformer, an electromagnetically induced current

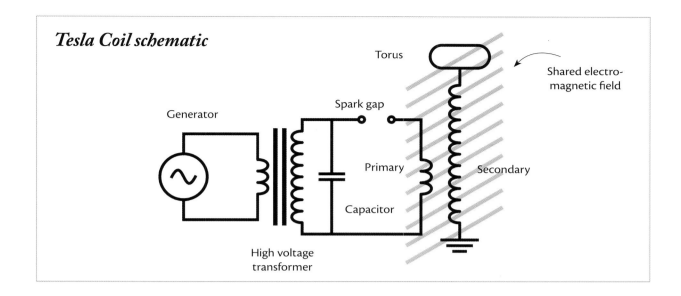

Tesla Coil schematic

Torus

Shared electro-magnetic field

Generator

Spark gap

Primary

Secondary

Capacitor

High voltage transformer

travels from the primary winding to the secondary winding through a shared magnetic field. Known as "close coupling," this sharing results from separate wires being tightly wound around a common iron core. When trying to achieve very high voltages, however, the insulation between windings can deteriorate, allowing current to flow between turns. Tesla's design solved this problem by separating the primary and secondary windings with a large space (known as a "spark gap") between the two. As a result, in Tesla's coil, the primary and secondary windings shared only about 10 to 20 percent of the same magnetic field. Nevertheless, it was enough to induce a current between the windings and had the advantage of permitting the coil to generate much higher voltages without melting the insulating material used to separate turns.

Capacitors

Next, Tesla connected a capacitor between the AC generator and the primary winding of the induction coil. In effect, his design diverted some of the current used to power the primary coil and stored it as an electrostatic charge that would build up in the capacitor. The rest of the power would flow as an alternating current through the primary winding, inducing a current in the secondary winding. The two currents had a tendency to oscillate at the same frequency because both windings would respond the same way to changes in the shared magnetic field created in the spark gap separating the two wires. In fact, the oscillations corresponded to a sine wave, where the peaks represented periods of maximum induction and the troughs represented periods of minimum induction.

Resonance

Like Hertz, Tesla reasoned that the waves generated by pulses of induction in the secondary winding were similar to vibrations and, as such, would be susceptible to resonance. In physics, as in music, resonance is the tendency

Capacitors: Bursting Batteries

Like batteries, capacitors are a kind of energy storage device. A battery stores electricity as electrochemical energy, which is released slowly and steadily as electrons move from the anode to the cathode during a redox reaction (see p. 68). Capacitors store electricity as electrostatic energy in an electrical field and can release all of it in one large burst.

A simple capacitor consists of two conductive plates separated by a non-conductive region (called the "dielectric"). As an electrical current is applied to the capacitor, equal and opposite charges develop on the sides of the conductive plates facing each other. When this charge builds up, it creates a potential difference (also known as "voltage") across the dielectric, generating an electrical field inside the capacitor. At some point—depending upon the insulating properties of the material separating the plates—the voltage exceeds the capacity of the dielectric separating these attracting forces, and, for an instant, the insulator behaves like a conductor, releasing all of the built-up energy as an electrostatic discharge.

It may be helpful to think of a capacitor like a flexible membrane stretched across a water pipe. As the flow of water (the current) backs up against the membrane, water pressure (voltage) increases. The more water is applied, the tighter the membrane stretches, and the more force it must exert to hold the water back. While the

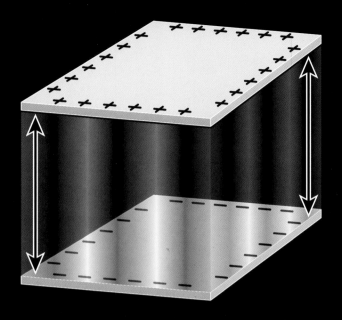

Illustration of a simple capacitor

membrane is being stretched, the pressure buildup (voltage) results from (and, thus, lags behind) the water flow (current). At some point, the membrane can't take the pressure. It breaks, releasing both the pressure and the water. It is important to note, however, that pressure builds up behind the membrane as a response to increased water flow. Once the capacitor discharges, however, pressure (voltage) decreases faster than the water floods (current). Thus, at discharge voltage leads current (for a moment).

of a vibration (like a sound wave) to oscillate with greater amplitude when a relatively small force is applied at the same frequency.‡ Perhaps the most well-known example of this phenomenon is when a soprano shatters a crystal wine glass with the mere force of her voice. In fact, depending on various factors (like the shape of the bowl and the thickness of the crystal), when force is applied to the glass, it will vibrate at a specific frequency. As the soprano belts a note of matching frequency, the sound waves she exerts on the glass introduce a relatively small force. This small force is reinforced by the physical properties of the glass, creating a positive feedback loop that will keep growing until the glass shatters (assuming the small input force is steadily applied).§

Tesla's oscillating transformer used a capacitor in much the same way. As an electrical charge builds up in a capacitor, it surges back and forth between the conductive plates until it is released as a high-frequency electrostatic discharge. By adjusting properties of the induction coil (like the length and thickness of the windings), the discharge from the capacitor could be timed to coincide with the maximum period of induction. In this way, the relatively small boost of current from the capacitor would match and reinforce the pulses of current induced by the magnetic field generated in the spark gap between the windings.[19]

Tesla's design also took advantage of the fact

Opera singer shattering glass

that when a capacitor is being charged, voltage builds up and thus lags in time behind current, like water pressure building up as water accumulates behind a membrane. Just the opposite is true during induction. Because an inductor presents a greater obstacle to a changing current than voltage, during induction current tends to lag behind voltage.

In Tesla's coil, the spark gap acts as a kind of insulator. As energy builds up in the capacitor, the voltage difference across the spark gap increases and the air within it becomes increasingly charged (ionized). Once the air reaches its limit, for a split second it acts like a conductor, allowing current to flow across the gap. After the current is released, the voltage across the gap equalizes and the air acts like an insulator again.

Here is where the difference between voltage leading or lagging current becomes critical.

---

‡ Technically, resonance also occurs at harmonic frequencies, which are integer multiples of the resonant frequency. But we'll keep things simple.

§ This parlor trick is hard to perform by an untrained singer, not only because of the high pitch and volume required, but because of the need to maintain a precise pitch to stay within the narrow frequency band of the glass's resonant response. Try singing a steady note into an electronic guitar tuner and you will soon discover just how difficult this is.

If the capacitor is timed to discharge just as voltage drops across the spark gap, current and voltage can be made to chase each other back and forth across the gap.[20] Voltage chases current, building up an electrostatic charge in the capacitor. When the capacitor discharges, it sends a current through the conductive wire chasing after voltage. If all of the physics seems confusing, the key concept to note is that introducing a capacitor into the configuration of an induction coil can generate an electrical oscillation. Moreover, because of resonance, if the oscillation created by this game of electrical tag is timed correctly, it will build upon itself. In Tesla's case, the result of this configuration was a coil that could generate higher frequencies and much higher voltages than any other device known at the time of its invention.[21]

Hardly Transformational

The Tesla coil may be one of the inventor's least understood inventions.[22] Confusion over the nature and utility of an "oscillating transformer" may, in part, explain why some erroneously claim that Tesla invented the electrical transformer that is so critical to AC power transmission.¶ Some responsibility, however, falls to the inventor himself. In his 1919 autobiographical essays, not only did Tesla write that his coil had come into "universal use and wrought a revolution," he also declared that the invention was "as revolutionary as gunpowder was in warfare."[23]

Tesla's coil—though capable of achieving higher frequencies than any previous

device—has remained more of an electrical oddity than a revolutionary tool. Leland I. Anderson, former president of the Institute of Radio Engineers, went so far as to declare that Tesla's contributions (like his coil) to the high frequency field "have been remarkably sterile."[24] In fact, while Tesla's oscillating transformer was able to achieve unprecedented frequencies and voltages, neither the transformer design it built upon, nor the concept of electrical resonance it exploited was revolutionary. Rather, as Tesla himself admitted, the coil and other inventions he patented during this time were mere evolutions of existing ideas, "nothing more than steps forward . . . to improve the present devices, without any special thought for far more imperative necessities."[25] A brief review of Tesla's travels reveals both the true origin of these revolutionary concepts, as well as the many opportunities the inventor had to capitalize on ideas that far preceded him.

First, it is important to note that, while the rudiments of Tesla's oscillating transformer initially arose in his patent for an "electrical transformer or induction device" (filed in March of 1890), the contraption we know as the Tesla coil evolved through multiple iterations, most of which are contained within patent filings stretching from March 1890 to July 1893. Tesla's original patent filing for an electrical transformer, for example, was the first to propose the idea of creating a large air gap between the primary and secondary windings.[26] However, Tesla did not raise the idea of replacing the interrupter in a standard induction coil with the electrostatic discharge from a capacitor until February 4, 1891, when he filed an application for a "method

¶ No doubt some of the confusion stems from Tesla's association with our modern AC power distribution system, which depends entirely on the use of electrical transformers (not Tesla coils).

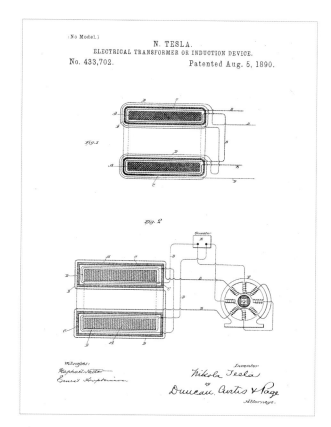

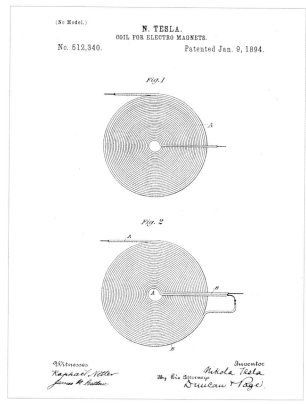

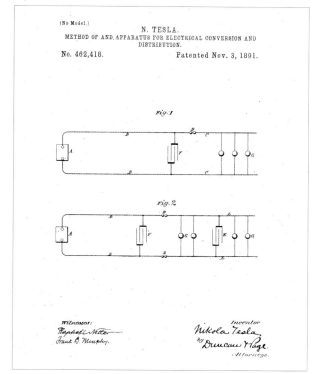

Patent drawings from patents making up Tesla's coil

and apparatus for electrical conversion and distribution."[27] Even then, he attributed this invention to "certain electrical phenomenon [*sic*] which have been observed by eminent scientists," but which had not yet "been utilized or applied with any practically useful results."[28] The idea of constructing windings of specific lengths to match the inductive discharge from a capacitor (the resonance component of Tesla's coil) does not even arise until Tesla's patent for a "coil for electromagnets," which was filed on July 7, 1893.[29]

During this same period, several important historical events in the development of transformers occurred. Most notably, in August 1891, Charles Brown and Mikhail

Dolivo-Dobrovolsky successfully transmitted AC power 175 kilometers, from Lauffen to Frankfurt, during a demonstration for the 1891 Frankfurt Electrotechnical Exhibition.[30] The pair accomplished this feat using a phased current (which closely resembled Tesla's polyphase system) and a transformer (then known simply as a "secondary generator") based on a design Lucien Gaulard and John Gibbs had first demonstrated in London in 1883.[31] According to some historians, George Westinghouse became aware of the Gaulard-Gibbs design from an article in the spring 1885 edition of the journal *Engineering*, which reviewed some the electrical innovations showcased at the 1885 International Inventions Exhibition held in London.[32] Upon reading about the innovation, Westinghouse promptly dispatched an agent to Europe to secure options on the Gaulard-Gibbs transformer.**

Not to be outdone, Westinghouse's main competitor, the Edison Company, soon negotiated its own options on a transformer system that had been designed years before Gaulard and Gibbs by a team of physicists working for the Ganz Works, a Budapest-based company that evolved from building ocean liners to constructing lighting systems (many of which, by 1885, were competing with Edison's systems for the bulk of the European electric lighting market).[33] In 1878, the company established a department of electrical engineering headed by a Hungarian engineer named Károly

Zipernowsky. Under his direction, Ganz entered the arc and incandescent lighting markets, installing systems throughout Europe.[34] Apparently, these systems utilized a crude type of transformer based on two copper wires wrapped around an iron ring.[35]

A remarkable young man named Ottó Titusz Bláthy began employment as a mechanical engineer for Ganz in 1883. Although Bláthy had virtually no formal education in electrical engineering, he learned much of the theory of electromagnetics from reading the articles of James Maxwell. After Bláthy observed the Gaulard-Gibbs AC power system at the Italian National Exhibition in Torino in 1884, he set about to make improvements on the design and was quickly reassigned to Zipernowsky's division. There, in the summer of 1884, he began to conduct experiments with the assistance of a young Serbian engineer named Miksa Déri. Together, the three engineers developed for the Ganz Works a new type of AC power distribution system based on the use of the first apparatus identified as a "transformer." The device consisted of two coils arranged uniformly around an iron ring and was intended to step down high-voltage AC power for use in incandescent lighting systems. The engineers first presented their system—which became known as the ZBD transformer—at the National Exhibition in Budapest in 1885.[36]

Prior Transformers

In his 2013 biography, Carlson recounted a story told by Serbian engineer Osana Mario at a conference held at the Technical Museum in Vienna in 1953 (Mario claims to have heard

** Westinghouse apparently paid Gaulard and Gibbs $50,000 for their system and had William Stanley install it in Great Barrington, Massachusetts, in 1886. Westinghouse tasked Stanley with using the new installation to develop an improved system that the industrialist could patent in the United States. Stanley's improved design—known as the Stanley transformer—became the prototype for the AC distribution system still used in the U.S. today.

it from Tesla himself when the two met at the 1893 Chicago World's Fair). According to Mario's account, Tesla began fiddling with a broken ring transformer he stumbled upon while either visiting or working at the Ganz Works in 1882.[37] Although this would have been prior to Bláthy's employment, and well before full development of the ZBD transformer, the period corresponds with the brief time Tesla spent in Budapest recovering from a bout of depression and taking evening walks with Anthony Szigeti. By early 1882, Tesla was working for the Puskás brothers, who transferred the inventor (and Szigeti) to Paris in April.[38] Nevertheless, it is conceivable that Tesla was exposed to an early version of the ZBD transformer while he was hanging around the Ganz Works in late 1881 and early 1882.

After Brown and Dolivo-Dobrovolsky's Frankfurt demonstration of AC power, Tesla became concerned about the priority status

ZBD transformer

of his patents and set out on an exhaustive European tour designed, in part, to put to rest any doubts about the novelty of his innovation.[39] Tesla apparently filed the first patent application on what would become the "Tesla coil" before he left for Europe. However, Seifer noted that the invention was not finalized until after Tesla returned to New York in late August 1892.[40] It is probable, therefore, that Tesla made important insights into the design of his coil over the course of his European tour. In any event, the inventor arrived in London on January 26, 1892, and immediately set about on his quest to quiet any competition by meeting with a reporter from the London-based journal *Electrical Engineer*. In a lengthy article published just three days later, Tesla made the case for how he developed his polyphase motor before Brown and Dolivo-Dobrovolsky (and, incidentally, prior to the publications of Galileo Ferraris).[41]

On February 3, 1892, Tesla presented his first European lecture before the London Institution of Electrical Engineers. Entitled "Experiments with Alternate Currents of High Potential and High Frequency," the talk highlighted Tesla's experiments with high-frequency currents and their interaction with electrostatic fields. Specifically, Tesla noted the work of the British physicist John Ambrose Fleming, who had suggested that vacuum tubes could be illuminated by exciting them with currents of specific wavelengths.[42] Fleming, who had lectured previously on electrical transformer theory at Cambridge and the University of Nottingham, was not unknown to the U.S. engineering community and was, in fact, hired as a consulting electrician to the Edison Company sometime shortly after Tesla's lecture.[43] Due to the overwhelming response to Tesla's presentation, he was asked to expand his remarks with an unplanned talk before the Royal Institution of London the very next day.

Using Capacitors

Fleming had been present at both lectures and invited Tesla to visit him at University College London over the weekend. At his laboratory, Fleming had successfully constructed an apparatus eerily similar to Tesla's oscillating transformer. In the letter in which Fleming made his formal invitation—dated February 5, 1892—the professor details how he had successfully harnessed "oscillating discharges with a Spottiswoode Coil as the primary and Leyden jars as the secondary."[44] A Spottiswoode coil was a very large induction coil—capable of producing up to 1.2 million volts—named after the British scientist William Spottiswoode, who invented the device in 1877. The primary coil of a Spottiswoode coil consisted of 1,344 turns of thin copper wire and was charged with a battery connected to a capacitor.[45] A Leyden jar was an early type of capacitor (then known as a "condenser") that stored electricity as an electrostatic charge between two electrodes, one on the inside of a glass jar, the other on the outside of the jar.[46] Apparently, by using a separate capacitor instead of a spark gap, Fleming's apparatus had produced the same kinds of current oscillations that Tesla would eventually achieve with his coil design. Although Fleming never applied for a patent on the apparatus he configured in his University College laboratory, the configuration did make its way into a 1905 patent Fleming filed on an "Instrument for Converting Alternating Electric Currents into Continuous Currents." A large portion of the decision the Supreme Court handed down in Marconi Wireless v. United States was concerned with how Tesla may have "borrowed" ideas from Fleming's 1905 patent when the Serbian filed his patents on components that would become essential to radio. Ultimately, the Court overturned Fleming's patent on the basis that a critical component—wholly separate from the oscillating configuration—was covered by a patent issued to Edison some twenty years prior.[47]

Although it is unclear whether Tesla accepted Fleming's invitation and saw the professor's apparatus for generating oscillating discharges in person,†† it is certain that the inventor left England for Paris sometime during the second week of February 1892 (giving him sufficient time to look in on Fleming).[48] In Paris, Tesla gave a well-documented lecture before a joint conference of the Société de Physique and the Société Internationale des Électriciens. He also received a telegram that his mother was dying and embarked on the long journey from France to his childhood home in Gospić, Croatia. Although the inventor arrived before her death, Djuka Tesla would not survive through the end of April. Tesla stayed on in Gospić for six weeks to help oversee the funeral arrangements and to reunite with members of his family he had not seen for over a decade.[49]

During this time, Tesla made several side trips, including at least one visit to the

---

†† According to Seifer, a collection of primary sources collected by the Nikola Tesla Museum in Belgrade (entitled, "Tribute to Nikola Tesla") contains evidence that Tesla did, indeed, meet with Fleming. See Seofer, *Wizard*, Bibliography: Frequently Cited Sources by or about Nikola Tesla, 479.

ABOVE: John Ambrose Fleming

LEFT: Spottiswoode coil

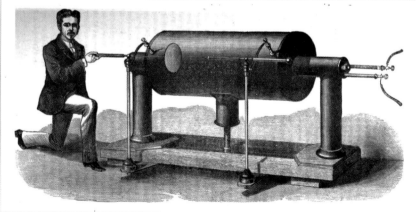

Ganz Works (renamed Ganz & Company) in Budapest. It is unclear exactly what Tesla did there, but the company was apparently in the middle of constructing a very large (1,000 horsepower) AC generator.[50] There is little doubt that such an undertaking would have been overseen by Zipernowsky, and it is not inconceivable that the inventor was exposed to work Zipernowsky, Bláthy, and Déri were engaged in during the summer of 1892, including the high-voltage transformers the Ganz team was developing as part of the largest European hydroelectric project yet undertaken by any company. The system Zipernowsky, Bláthy, and Déri had designed was capable of transmitting AC power from high-voltage generators in Tivoli to a distribution system in Rome, where the voltage would be stepped down for use by individual consumers.[51]

Tapping Resonance

On his return to Paris, Tesla made arrangements to stop at the University of Berlin to visit the renowned German physicist Hermann Ludwig von Helmholtz, who had been studying Maxwell's theories on electromagnetic radiation, especially with regard to the nature of light and color.[52] Specifically, Helmholtz was busy formulating an electrodynamic theory of action at a distance that theorized that electric and magnetic forces propagate instantaneously.[53]

From there, Tesla traveled to Bonn to visit Heinrich Hertz, who had completed his postdoctoral studies under Helmholtz (and served as his lab assistant until 1883). It had been only five years since Hertz had astounded the electrical engineering community by proving Maxwell's theories on the nature of electromagnetism using instruments, like his resonant wire loop, designed to transmit and receive radio signals. Hertz's instruments bore some startling resemblances to components that Tesla would add to his coil design. As part of his experiments with ultra-high frequency electromagnetic waves, for example, Hertz had generated radio waves from oscillating currents triggered by a high-voltage electrical discharge from a capacitor. His method of detecting

the waves, moreover, involved an antenna connected to a spark gap. In simulating these experiments later for lecture purposes, Hertz found that he could study the high-frequency waves by observing periodic discharges from the primary winding of an induction coil as they jumped across a spark gap, a setup uncannily similar to the coil Tesla would eventually design.[54]

Although Tesla may have been first to the patent office, it is probable that Hertz had worked out the resonance of electrical oscillations and even invented a mechanism to exploit them to produce high-frequency waves well before Tesla completed his patent filings. In his exhaustive 1998 biography, Seifer noted that Hertz displayed the resonance effects between primary and secondary windings to

Hermann Ludwig von Helmholtz

Tesla using the looped wire apparatus, and had—well before Tesla completed the design of his coil—established the existence of "standing waves" that occur when the alternations of power charging the primary winding correspond with the oscillations induced in the secondary winding.[55] Although the biographer mentions that Tesla left Hertz on bad terms, Seifer never appears to recognize the more significant implication: that Hertz's concepts of resonance (and his laboratory set-up) preceded Tesla's patents on his famous coil.

While it is clear that Tesla was the first to patent an induction coil that utilized large air gaps between the windings, added capacitors to generate high-voltage oscillating currents, and synchronized the discharge of the capacitors to generate resonant frequencies, it is also clear that many of his patents were not filed (or were substantially modified) until after he had been exposed either to the underlying concept or a strikingly similar apparatus (or both) during his European tour of 1891–1892. For example, while he was exposed to Hertz's concept of electrical resonance (and very likely to Hertz's own devices for generating resonant frequencies) in late summer of 1892, these concepts did not make their way into Tesla's patent filings until 1893, after he had returned from his tour. Nevertheless, partly on the basis of lectures he gave during those travels, Tesla would claim credit for announcing the invention and his name would forever be associated with the apparatus.

"The constant appearance of a name in connection with the development of a given art, science, discovery or invention makes an impression which is difficult to destroy, and this is true even among the most intelligent classes."

—T. COMMERFORD MARTIN, *SCIENCE*, NOVEMBER 2, 1900

chapter six

Wireless Transmission

After World War II, the German engineering genius (and former Nazi SS officer) Werner von Braun was brought to the United States to advance ballistic missile technology under the U.S military's top-secret Operation Paperclip. Eventually, von Braun would oversee development of NASA's Saturn V launch vehicle, the rocket responsible for getting the Americans to the moon and back. One afternoon in 1967— just after the engineer was appointed director of NASA's Marshall Space Flight Center— he found himself in the Center's conference room about to witness a demonstration by a team of scientists from Raytheon's Microwave & Power Tube Division.[1]

For this momentary break from the moon-shot frenzy, von Braun could thank a Raytheon electrical engineer named William C. Brown. Brown had made headlines three years earlier when his team publicly demonstrated that they could keep a small, unmanned helicopter

aloft for ten hours powered solely by a beam of microwaves transmitted from the ground.[2] Now, Brown was standing at one end of a long conference table about to show the world's foremost rocket scientist how he did it. In one hand, he held an antenna attached to a small motorized fan. At the other end of the table sat a parabolic reflector about three feet in diameter attached to a microwave tube. When the electrical switch was thrown, the fan began to spin like the propeller of an airplane. Brown waved his hand several times between the reflector and the antenna to prove that the energy powering the motor was coming from a 100 watt microwave beam. Although the theory behind wireless transmission of electricity arguably originated as early as 1868, Brown and his team had overcome one of the greatest challenges to its practical application: concentrating electromagnetic energy in a beam narrow enough for efficient point-to-point transmission.[3] Von Braun (who was more accustomed to impressing than being

OPPOSITE: NBS wireless phone demonstration, 1906

impressed) was awed enough that he contracted Raytheon to study how the technology could be expanded to power a satellite.[4]

Today, Tesla's name has become almost synonymous with the dream of wireless transmission. Some claim it was his most important (if unrealized) invention.[5] In fact, many of Tesla's most ardent admirers await the day some enterprising company perfects his technology and vindicates the inventor's genius by implementing a large scale system of power distribution devoid of wires.[6] To appreciate the vision (and challenge) of wireless transmission, as well as Tesla's contribution to it, will first require, however, a basic understanding of the science of electromagnetic resonance.

Understanding Electromagnetic Resonance

In 1868, three years after he had stunned the world of physics with his groundbreaking theories on electromagnetic radiation, James Maxwell spent an evening with Sir William Robert Grove, a Welsh judge and scientist (who invented the first incandescent electric light nearly forty years before Edison perfected his bulb).[7] Earlier that day, Grove had been busily engaged in experiments on electrical discharges through vacuum tubes. He found that when he connected a capacitor to the primary winding of an induction coil powered by an AC generator, he could produce much larger sparks (indicating, of course, much larger current flowing through the primary coil) than without the capacitor. Perplexed, and knowing of Maxwell's mathematical genius, Grove challenged the young physicist to determine how interposing a capacitor between the generator and the spark was

producing currents many magnitudes greater than those supplied by the generator itself.[8]

Maxwell returned home and spent the entire night contemplating the problem and making calculations. In the morning, he drafted a letter to Grove with the answer.* Maxwell surmised that the capacitor discharged its stored electrostatic energy in the form of a vibration at a frequency proportional to the alternation of the current powering the primary wire, a remarkably accurate explanation of electrical resonance (though Maxwell never actually used that term) some twenty-two years before Tesla constructed his oscillating transformer.[9] By producing a discharge at a corresponding frequency, the capacitor could use resonance to generate an alternating current far greater than the input (see page 123).

Electrical resonance, like all resonance, is a characteristic of waves (in this case electromagnetic waves). Any electrical current or any changing magnetic field generates electromagnetic radiation, radiant energy that travels outward through space from the source. Sometimes this energy acts like a wave. Recall that electrical currents are associated with electric and magnetic fields. A changing magnetic field, moreover, induces an electrical current…which generates a magnetic field, which induces an electrical current (and so on). An electric or magnetic field, in other words, is induced by changes in the other type of field. The right-hand rule tells us that electric fields point at right angles to the magnetic fields they induce, or that induced them. Moreover, this repeated interaction between the two

---

* The judge quickly dispatched it to *The Philosophical Magazine*, which published it in May, 1868.

kinds of fields then propagates in space, in a direction again perpendicular to the directions in which the alternating electric and magnetic fields are pointing. Therefore, we can imagine electromagnetic radiation as a *transverse wave* of oscillating electric and magnetic fields.

Electromagnetic waves have the same general characteristics as other waves. The distance between crests of maximum output determine wavelength. The size of those crests (measured from neutral to maximum) determines the amplitude. Frequency is the rate of oscillation within a given unit of time. Named for the German physicist who proved Maxwell's famous theory, a Hertz is the unit of measurement of a wave's frequency and corresponds to one wave cycle per second. It is important to remember that, while all electromagnetic radiation propagates through space at the same speed—the speed of light, or about 300 million meters per second—the frequency of electromagnetic waves can change depending on the oscillations inducing them. Hence, all electromagnetic radiation falls within a spectrum characterizing the

frequency of its waves. Radio waves, for example, typically have frequencies ranging between 3 kilohertz (3,000 wave cycles per second) to 300 gigahertz (300 billion wave cycles per second).

It is easy to see that waves with high frequencies must necessarily have shorter wavelengths and waves with low frequencies have longer wavelengths. In other words, the frequency of an electromagnetic wave is inversely proportional to wavelength. As the length of a wave increases, its frequency decreases, and vice-versa. The spectrum of electromagnetic radiation, therefore, can just as well be divided according to wavelength—from large waves (smaller frequencies) to small waves (higher frequencies)—into sound waves, radio waves, microwaves, infrared radiation, visible light, ultraviolet radiation. X-rays and gamma rays.

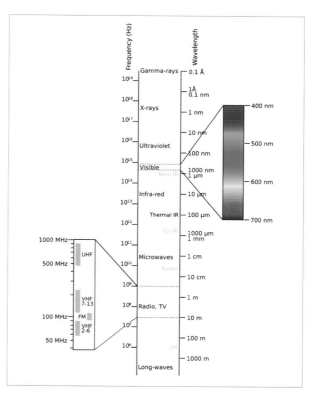

Electromagnetic spectrum

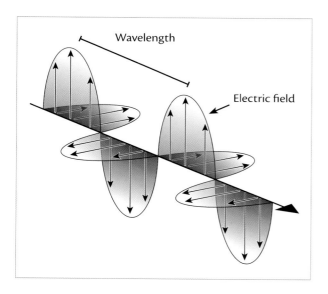

Frequency, moreover, is directly proportional to energy. The higher the frequency, the more energy the wave carries. Like the visible light spectrum, the boundaries between different types of electromagnetic radiation are not distinct, but fade into each other. You might wonder, therefore, why physicists have drawn lines at particular frequencies to characterize different types of radiation. The answer has to do with the fact that, even though it often acts like a wave, just like light, electromagnetic radiation sometimes behaves like a particle. The truth is that the vibratory energy that comprises all radiation (including light) moves in tiny packets called photons. As an atom absorbs a photon, the electrons within it become excited, which causes them to jump to an orbit farther away from the atom's nucleus. This jump is often referred to as a "quantum leap." When energy is transferred by radiation, the excited electrons descend back to a lower-energy orbit and emit a photon of energy in the process. This game of sub-atomic hot potato allows energy to pass from molecule to molecule (in a wave-like fashion).

When an atom is just sitting there minding its own business, it is said to be in a particular "quantum state," literally the energy condition of the atom. After being jostled by an incoming photon, the atom takes on a different quantum state. When it emits a photon, the atom returns to the previous quantum state. The difference between the two quantum states determines the frequency of the emitted photon. However, the atoms of any given element exist in discrete states (unique to that element) that generally correspond to the type of electron orbits generated by the atomic structure of the element. That is, because electrons orbit

elements in different ways (and at different distances), the frequency of photons emitted from one element will be different from the frequency of photons emitted from any other element.[†] As a result, through a process called "spectroscopy," scientists can observe the frequencies of photons released from a substance to determine what elements it contains. Astronomers, for example, can use this technique to determine the composition of stars that are billions of light years away.

Historically, the manner in which photons of electromagnetic radiation interact with different kinds of matter (and, hence, their different frequencies) was used to characterize different types of radiation. Thus, ultraviolet radiation roughly describes the frequencies of photons with enough energy to excite the *outer* electrons of many elements to the point that they escape the atom altogether (a process known as the "photoelectric effect"). X-rays roughly describe the frequency of photons with enough energy to cause the core electrons of an atom to fly away. (The damage this causes to the atom is one reason these forms of electromagnetic radiation are so dangerous to living organisms.)

When electricity is transmitted through a conductive wire, engineers generally are more concerned with what is going on inside the conductor (resistance, thermal expansion, etc.). While the current running through the

---

[†] An (imperfect) analogy would be to think of electrons as moons orbiting a planet. Saturn's sixty-two known moons revolve in orbits determined by their mass and the planet's mass (which generates gravitational attraction between the two), as well as the gravitational pushes and pulls of the planet's other moons. As a result, their orbits differ entirely from those of Jupiter's sixty-seven confirmed moons. Similarly, the orbits of electrons correspond to interactions—like the electron's energy and its angular momentum—between the electron, the atom's nucleus and all of the atom's other electrons.

The author Francis Marion Crawford observes a spark gap in Tesla's Liberty Street laboratory

wire still creates a magnetic field, it is simply a by-product, not important to a utility's transmission of energy (though these EMF fields may be very important to the people who live near high-voltage lines).

Wireless transmission—of anything from radio waves to microwaves—on the other hand, is entirely concerned with the electromagnetic field external to the conductor. The waves these fields generate propagate through space like an expanding balloon. Unlike a balloon, however, the strength of these waves decreases with the square of the distance from their source.‡ Because electromagnetic radiation dissipates in all directions, therefore, wireless transmission is generally unsuitable for transmitting electrical power, at least at great distances.[10] The strength of the wave of electromagnetic energy simply dissipates too quickly to be of much use.

Tesla's Ground Game

For Tesla, wireless transmission presented a challenge he was convinced he could overcome. He had scarcely returned from his 1892–1893 European tour when he accepted an invitation to give a lecture on February 25, 1893, to the National Electric Light Association in St. Louis. He captivated the standing-room-only crowd when he waved glowing fluorescent tubes like light sabers, powered solely by the strong electromagnetic field he created with his

induction coil. After several startled spectators fled the Exhibition Theater accusing Tesla of doing "the devil's work," for the first time that we know of, the inventor mentioned the idea of wireless transmission:

*Some enthusiasts have expressed their belief that telephony to any distance by induction through the air is possible. I cannot stretch my imagination so far, but I do firmly believe that it is practicable to disturb by means of powerful machines the electrostatic condition of the earth, and thus transmit intelligible signals **or perhaps power**.*[11]

Tesla claimed that he had prepared a detailed outline of his vision for wireless transmission of power, but omitted it from his remarks that night, fearing his bold suppositions might scare off potential investors.[12]

Despite Tesla's exceptional ability to imagine whole contraptions (and the worlds they occupy) in his head, he dismissed summarily the idea of wireless transmission through the air, at least initially. Most of the "enthusiasts" the inventor referenced were examining the omnidirectional transmission of radio waves from a grounded antenna to a transmitter located some distance away. This model of a wireless circuit envisions a primary current (signal) transmitted from an antenna with a return current traveling from the receiver, through the earth, back to the antenna. Tesla was convinced that this system—with electromagnetic waves radiating from the source in all directions— would be hopelessly inefficient as a method for transmitting electricity.[13] Instead, he decided to turn the system on its head by using the earth as a conductor to send the primary current from

---

‡   The "inverse-square law" of electromagnetic radiation is the result of the spherical boundary created by the propagation of the waves themselves. Archimedes first calculated that the surface area of a sphere is $4 \times \pi \times r^2$ (the square of the radius). The radius of a sphere of expanding electromagnetic energy, therefore, is the distance from the source of the radiation. Since the energy transmitted by any given electromagnetic wave is distributed evenly across its expanding spherical boundary, the amount of energy at any given point is determined by r^2, the square of the distance from the power source.

a grounded transmitter, to a grounded receiver, with the return current sent through the air from an antenna back to the transmitter.

Although it is easy for some historians to characterize Tesla's ruminations about ground electricity as "thinking like a maverick" (given that contemporaries like Hertz and Marconi were focused on transmitting electricity through the air), the fact is that the concept of electrical grounding was nothing new even in Tesla's time.[14] Since about 1820, most long-distance telegraph systems used at least two wires, one to carry the signal and the other to carry a return and complete the circuit. However, sometime in late 1836 or early 1837, the German physicist Carl August von Steinheil found that he could eliminate the return wire entirely if he connected one end of the transmitting wire to a metal plate buried in the ground.[15] Thus, even while Tesla

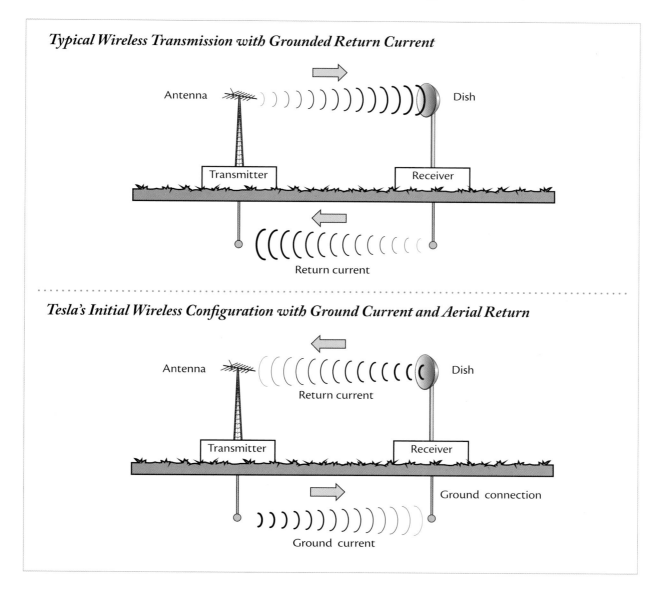

Typical Wireless Transmission with Grounded Return Current

Antenna   Dish
Transmitter   Receiver
Return current

Tesla's Initial Wireless Configuration with Ground Current and Aerial Return

Antenna   Dish
Return current
Transmitter   Receiver
Ground connection
Ground current

Three Ways to Transfer Thermal Energy

In simple terms, there are three different ways heat energy can be transferred: conduction, convection, and radiation. Conduction transfers energy between adjacent atoms that "vibrate" against one another, creating a kind of domino effect. Because it requires atoms to be packed close together, conduction generally works best in solid materials, like metal wires.

Convection transfers energy as atoms in a fluid state flow past one another. As they slip around, energy passes from hotter places to colder places in an attempt to reach equilibrium. Convection does not require that atoms be densely packed together, as in a solid state, but is generally how energy is transferred within liquids and gases.

Radiation transfers energy by passing it between atoms in tiny packets. As an atom becomes excited, it releases a packet of energy that travels at light speed until it happens to hit another atom. The receiving atom absorbs the energy, becomes excited, and releases it again, in a continuous game of atomic hot potato. Energy transmission by radiation does not require atoms to be located anywhere near one another and can occur through the vastness of (mostly) empty space. This is one reason that spectroscopy of electromagnetic radiation can tell astronomers a lot about the characteristics of stars thousands of light-years away.

was thinking about using ground currents for wireless transmission of electricity as early as 1893, speculation had existed for decades about using ground currents for wireless telegraphy.[16]

Upon his return to New York City, Tesla began to experiment, sending high-frequency pulses into the ground, and generating streams of electricity from the earth to a return terminal hanging in the air.[17] Interestingly, Tesla rejected the common belief that the return current derived from waves of electromagnetic energy radiating in all directions. Rather, he maintained that it resulted from electrical oscillations moving, by conduction, through gas particles at low pressure in the upper atmosphere. Tesla knew that the air at sea level tends to act like an insulator (as, for example, the air between a spark gap). However, the inventor had observed how noble gases at low pressures (as in Geissler tubes), glowed when excited by a high-voltage current. He reasoned, therefore, that the higher the voltage and the lower the atmospheric pressure, the more power could be transmitted through the air.[18]

In his Fifth Avenue laboratory in early 1894, Tesla began to test ways he could distribute high-voltage currents through the air to power lights within the room. In his wired set-up, the inventor connected his induction coil to a step-up transformer. He soon discovered, however, that he could use his oscillating transformer to generate power at high enough voltages that he could transmit it wirelessly between a primary winding at one end of the room and a secondary winding at the other end.[19] Essentially, he turned his entire laboratory into a giant spark gap, where the air in the room was saturated with a magnetic field shared between two coils placed at distances around the room. He would use his Tesla coil to charge a bank of capacitors that would discharge high-voltage currents into a thick wire that ran around the circumference of the room. This wire acted like the primary winding of a transformer, only with a single, large turn (the room's circumference). As a secondary winding, Tesla used a coil of wire three feet tall mounted on castors (allowing him to test how it received the wireless current at different locations around the lab).[20]

By February of 1894, Tesla was demonstrating his new wireless lighting system to friends and luminaries (and even a few awestruck reporters) in what Carlson described as a kind of elaborate self-promotion designed to attract investors.[21] This was a curious strategy, given that Tesla had not yet patented the system, and

Mark Twain in Tesla's lab with glowing tubes

that he was still working with Page and Brown, who had previously admonished the inventor to keep quiet about any of his new discoveries. In this case, Tesla did just the opposite of what Page and Brown instructed, interviewing for a string of laudatory articles and helping to distribute a plethora of provocative photographs (mostly of notable visitors to his lab that he mesmerized with glowing tubes of fluorescent gas).

By February 1895, Tesla convinced Edward Dean Adams (champion of the Niagara hydro-electric project) to join him and Alfred Brown in forming a new commercial venture—the Nikola Tesla Company—that would promote Tesla's new high-frequency innovations. But, like so many of his inventions, Tesla failed to turn his dramatic demonstrations into workable commercial products. By late 1895, the company failed to attract a single notable investor.[22]

Rather than critically examine his repeated failure to commercialize his inventions, Tesla decided that the key to attracting investors was to go bigger. He expanded his focus from wireless lighting to the vision of a worldwide system of wireless power he had first hinted at during his 1893 lecture. He immersed himself in the literature of electrical oscillation and resonance. Knowing that the earth carried electrical charges, the inventor postulated that he could use the planet as a giant transmitter of electrical energy.[23] If he could somehow measure the characteristic frequency of the earth as an electrical conductor, Tesla thought he might be able to pump currents into the earth at resonant frequencies to boost the output at any grounded receiver. In late 1894 and early 1895, he undertook a number of clandestine experiments around New York City, some of which he described in later writings. Before he worked out all the details of his wireless system, however, his lab burned to the ground along with (almost) all of his testing equipment.

It took Tesla two years (and a serious bout of depression) to get over the loss of his lab. During that period, he faced mounting monetary challenges, including paying for lost equipment that he had forgotten to insure. Scraping together what funds he could, the inventor rented space at 46 East Houston Street and began to rebuild the laboratory. From the summer of 1895 through 1896, he worked on electrical oscillations and resonance, filing patents for a redesigned (smaller) oscillating transformer that, in combination with a new vacuum bulb he invented, could provide lighting more efficiently (and at less cost) than Edison's incandescent system.[24] During this time he also explored X-rays (which had been discovered by the German physicist Wilhelm Conrad Roentgen) and rebuilt a number of prototypes of remote controlled devices he claimed had been destroyed in the 1895 fire.[25]

On September 2, 1897, Tesla filed the first of his patents on a "system of transmission of electrical energy."[26] Although it would become the basis of Tesla's claim to the invention of radio, the inventor made clear that his intention was to generate an electrical current "to traverse elevated strata of the air between the point of generation and a distant point at which energy is to be received and utilized."[27] Essentially, Tesla proposed an enlarged version of the set-up he had used to demonstrate wireless lighting in his laboratory: a step-up transformer in which the common magnetic core was the distance between an elevated terminal of the primary winding and an elevated terminal of the secondary winding. By creating a massive difference in electrical potential between primary and secondary windings (a voltage difference) separated by low-pressure air, Tesla reasoned that, like the flash of a spark gap, "a current will be transmitted through the elevated strata, which will encounter little and possibly even less resistance than if conveyed through a copper wire."[28]

It is important to note that Tesla's 1897 patent application was more for a method than an apparatus. While he had demonstrated the rudiments of his system of wireless transmission in the laboratory, he had not fully tested whether a working prototype could actually power electrical devices hundreds of miles away. As a result, the description Tesla included for a device capable of implementing his method was necessarily vague. His 1897 patent described, for example, "a coil, generally of many turns of a very large diameter, wound in spiral form about a magnetic core or not, as may be found necessary."[29]

Read carefully, Tesla's 1897 application (filed before his Colorado experiments) demonstrated that Tesla did not yet understand or appreciate how electrical resonance would play into his system. Although the inventor noted that achieving the great voltage differences his method required could "be facilitated by using a primary current of very considerable frequency," he pointedly noted that "the frequency of the current is in large measure arbitrary," so long as the voltage difference between the terminals was sufficiently high.[30] Nor, incidentally, had Tesla embraced the increasingly accepted theory of electromagnetic radiation, instead repeatedly describing his process as "conduction" between gaseous particles.[31]

While Tesla was toying with wireless transmission (and dreaming of a worldwide system of wireless power), Guglielmo Marconi had far less ambitious goals. After Hertz went public with his breakthrough in electromagnetic radiation, Marconi began experimenting with

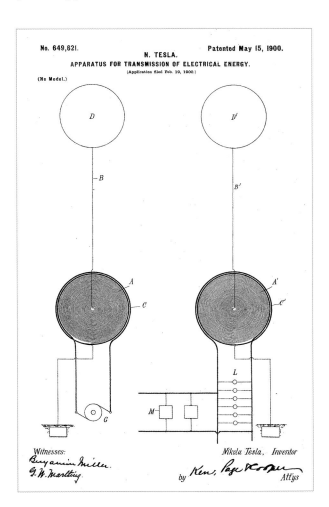

Patent drawings from Patent No. 649,621 "Apparatus for Transmission of Electrical Energy"

electromagnetic radiation for wireless telegraphy, incrementally improving his devices until, by late 1898, he declared that he could send messages wirelessly up to one hundred miles away.[32] Marconi's boasts were a personal affront to Tesla, who felt that the Italian was simply running experiments based on theories Tesla himself had articulated in several lectures dating back to 1893. Once Marconi successfully (he claimed) transmitted signals across the English Channel in March 1899, Tesla felt compelled to trump him with his own claim to the *New York Journal* that he was perfecting a system that would be capable of sending "a 2,000 word dispatch from New York to London, Paris, Vienna, Constantinople,§ Bombay, Singapore, Tokio [*sic*] or Manila."[33]

Tesla had made such boasts before, with few practical results. Among his scientific peers, his fantastical claims increasingly were met with cynicism.¶ So, with Astor's money in hand (and spurred by his competition with Marconi), in early 1899 Tesla set out for Colorado, convinced he could discover the nature of ground currents, calculate the planet's electrical frequency, and somehow make good on his promise of a worldwide system of wireless telegraphy.

Colorado Confidential

What exactly happened in Colorado? The conventional story, replete with photos of Tesla's secretive (and windowless) Colorado Springs experimental station, abounds with mystery. Mystery breeds speculation, which fuels mythology. Popular accounts, for example, tell of Tesla lighting whole fields of 50-watt incandescent bulbs screwed into the earth, some twenty-six miles from the transmitting station.[34] The mystery surrounding Tesla's Colorado experience is odd, however, since (for perhaps the first time in his scientific career) the inventor kept a detailed diary of his time there. This fact makes the speculation about fields of wireless light bulbs even more bizarre. Tesla, for example, wrote only that he had successfully transmitted a ground current *of sufficient strength* to wirelessly light more than two hundred bulbs (though he admitted in later lectures that he had not "as yet effected a considerable amount of energy, such as would be of industrial importance" using the same method).[35] Nevertheless, the myth of the field of wireless incandescent bulbs has become part of popular lore. (For instance, the scene is played out in a fog-drenched mountainside, to a swelling music score, in the 2006 thriller *The Prestige*, starring Christian Bale and Hugh Jackman).

Through various experiments (duly noted in his diary), many of which failed entirely, Tesla nevertheless learned much about electrical resonance. By observing the frequent lightning storms on the Colorado plains, eventually he softened his position on whether wireless currents could be transmitted through the air, acknowledging that "the air strata at very moderate altitudes, which are easily accessible, offer, to all experimental evidence, a perfect conducting path."[36] As his language demonstrates, however, he still doggedly insisted that these

---

§   Although the city was commonly known as Istanbul even before the Ottoman conquest of 1453, it wasn't until 1930, when the Turkish Postal Service Law officially requested that foreigners refer to the city by its Turkish name that the name "Constantinople" fell out of fashion among western elites like Tesla.

¶   Amos Dolbear, a noted professor of physics and astronomy at Tufts University, proclaimed that so few of Tesla's startling announcements had come to fruition that the inventor was in danger of becoming, "like the man who called 'Wolf! wolf!' until no one listened to him."

currents resulted from conduction between molecules at low pressure, rather than through electromagnetic radiation.

As for ground currents, Tesla began to believe that pumping an oscillating electrical current into the earth at just the right frequency would create pulsations of energy, traveling in waves that expand outward over the planet's surface. These waves would grow in intensity until they converged on a point on the globe directly opposite their source. From this antipode, Tesla surmised, the waves would rebound and, because they were oscillating at the same frequency, intersect precisely with the crests and troughs of the out-going waves.[37] This superposition of opposing waves, each with the same frequency, would create what is known as a standing wave.

Standing Waves

In electrical transmission, a standing wave is a wave of electromagnetic energy that remains in constant position. It is formed any time current, voltage or electromagnetic field strength of the same frequency move in opposite directions. A more familiar example of this phenomenon is to imagine two people shaking the ends of a jump rope. If the two can synchronize their shaking at just the right tempo, onlookers would observe what appears to be a stationary wave of rope hanging in space. In practice, however, it is nearly impossible to generate a pure standing wave, since the reflected wave runs into materials with different resistance (known today as "signal distortion"). Instead, what power system operators

A scene from *The Prestige*

and radio technicians usually create are partial standing waves.

For Tesla, the inability to create a perfect standing wave was fortunate, since he believed a perfectly resonant wave would bounce back and forth, constantly reinforcing itself until, like an opera singer shattering a wine glass, the oscillation would become so powerful it would split the earth. Instead, the inventor envisioned a system of resonant transmitters, spaced across the globe, each pumping resonant currents into the earth, creating partial standing waves of electromagnetic energy, which could be tapped anywhere on the planet with a simple device (essentially a tall metal rod, a ground connection, and a tuning mechanism).[38]

Although Tesla spent the remainder of his time in Colorado trying to work out both the method and apparatus for realizing his wireless vision, his notes reveal that he routinely searched for evidence to confirm his theories and ignored evidence to the contrary. Carlson noted, for example, that the inventor never directly measured how far he could actually transmit useful amounts of power, but assumed that if his theory were proven at any distance, it was confirmed for all distances.[39] Despite the lack of experimental evidence for his theories, Tesla claimed that, "careful tests and measurements" in Colorado, "have demonstrated that power in any desired amount can be conveyed, clear across the globe if necessary, with loss not exceeding a few percent."[40] So upon his return to New York in mid-January 1900, the inventor was determined to raise the capital needed to realize his vision of a worldwide wireless communication and power system.

Hidden History

In his autobiographical essays, Tesla claims that, about the time of the fire that destroyed his Fifth Avenue laboratory, he had the first inklings that resonance would be the key to achieving higher voltages with his wireless system. Although he realized that he could simply use larger induction coils, the inventor later claimed that he "had an instinctive perception" that he could make "a comparatively small and compact transformer" generate disproportionately large "electromotive forces," although he was not yet certain of the mechanism.[41] This repeated allusion to instinctive perceptions, gut feelings that had yet to be confirmed through hard science, is one reason that Tesla's own accounts of his process of invention helped fuel the myth of the lone inventor, whose ideas spring *sua sponte* from the genius mind.

Although Tesla filed his first patent application for a wireless apparatus in 1897, it was not until he returned from Colorado that he added electrical resonance to his design. On February 19, 1900, Tesla filed an amendment to his 1897 patent application, expanding greatly the concept of electrical resonance and introducing for the first time the concept of standing waves.[42] Gone was the assertion that "the frequency of the current is in large measure arbitrary." In its place, Tesla emphasized the need for the transmitting coil and receiving coil to be synchronized in order to obtain "the best conditions for resonance."[43]

Historians appear to have overlooked that Tesla made these new (and critical) specifications as an amendment to his original application, erroneously recording the date of his basic radio patents as September 2, 1897.[44]

Who Invented the Telephone?

en the most unschooled person knows the classic
ory of Alexander Graham Bell, the Scottish-
anadian inventor (and U.S. citizen), who, in 1876,
buted the first documented telephone transmission
his assistant, Thomas Watson: "Mr. Watson, come
re! I need to see you." Bell was in an adjacent room,
uggling with his invention, when he spilled battery
d on himself and made his famous cry.[45] (At least,
at's according to Watson, who recorded the story
his memoirs written nearly a half-century after
e fact).[46] In reality, the history of the telephone's
vention involves dozens of colorful historical
aracters and more than 600 separate court cases.

hann Phillip Reis (1860)

schoolmaster and self-taught scientist, the German
ventor Johann Philipp Reis initially was heralded by
any European scientists as the first inventor of the
ephone.[47] Some still credit him with the achievement.
1859, Reis documented the results of experiments he
d conducted proving that electricity, like light, can
opagate through space without the need for wires.[48]
e next year, he constructed a device in which sound
ves struck a membrane, which actuated a platinum
er that would open and close an electrical circuit as
ounced on a contact made of platinum foil. Reis
imed that his "telephon" could transmit sound along a
re conductor for a distance up to 100 meters.[49]

There is some evidence that Reis lectured on his
ephon as early as 1854. What is known for certain
that he outlined how the invention worked in detail
ring a speech before the Physical Society of Frankfurt
October 26, 1861.[50] The next year, he demonstrated
improved model to the Inspector of the Royal
ussian Telegraph Corporation, who showed little
terest.[51] However, models of his device were sent to
ondon, Dublin and elsewhere.

Reis's telephon aroused some controversy when it
was introduced into evidence in a series of patent cases in
which Bell would ultimately prevail.[52] In the courtroom,
Reis's telephon could be made to transmit sounds,
but not intelligible speech.[53] Rather than support the
contention that Reis's telephon pre-dated Bell's, this
apparent failure—demonstrated by purported electrical
"experts," not the inventor himself—was seen by the
court as evidence of Bell's singular achievement. No
one appears to have invited Reis to the courtroom, even
though he had successfully demonstrated in 1861 that
his device could reproduce speech when he transmitted
(distinctly) the phrase "The horse does not eat cucumber
salad."[54] Sadly, however, no documentation of his
demonstration found its way into the legal record.

In an interesting historical footnote, in 1947,
engineers from STC (a British telephone company),

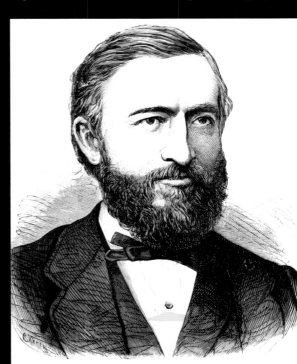

Johann Phillip Reis

made a few minor adjustments to a Reis prototype and found that it was able to reproduce speech "of good quality, but of low efficiency." However, at the time of their discovery, STC was in negotiations with AT&T (formerly the American Bell Telephone Company). Fearing that evidence undermining Bell's claim to the invention would disrupt negotiations, STC's chairman ordered that the tests be kept secret.[55] It was years later, when documents were found at the London Science Museum, before anyone outside the company knew of the successful tests of Reis's device.[56]

Amos Dolbear (1865)

The American physicist Amos Dolbear would gain fame for patenting a "mode of electrical communication" in 1886 that blocked Guglielmo Marconi's total monopoly on radio in the United States.[57] While a student at Ohio Wesleyan University in 1865 (some 11 years before Bell), however, Dolbear invented a "talking telegraph"

Amos Dolbear

with a receiver made from a permanent magnet and diaphragm made of tin.[58] At some point after he graduated, he reconstructed the device and ran a connection from his house at 70 Washington Street in Boston to a furniture warehouse (owned by one Francis Holmes) about ten blocks away.[59] When the American Bell Telephone Company started to assert Bell's exclusive license on the telephone, it brought a patent infringement suit against Dolbear and Holmes—as well as a dozen other defendants who were disclaiming Bell's priority—in the Federal Circuit Court in Massachusetts.

While the court ultimately upheld Bell's patent (and forced Dolbear to dismantle his connection), it only did so on a technicality. Judge Gray, writing for a panel of himself and Judge Lowell, found that, even if Dolbear had built a working telephonic *apparatus* prior to Bell, nowhere in his legal defense did he deny the patent granted to Bell for a specific *method* of transmitting the human voice over a conductive wire. In an absurd contortion of legal reasoning, in other words, the court held that Bell was the first to patent the telephonic process, even if Dolbear constructed a practical device utilizing the process years before Bell was granted the patent.[61]

Despite his legal defeat, the editors of *Scientific American* extolled the "important advantages" Dolbear's system had over Bell's method, not the least of which was that, "the words and voice of the speaker come clearly to the ear without the bubbling, crackling, sputtering, and whizzing noises that so seriously curtail the Bell system."[62] But Dolbear, like so many innovators lost to history, failed to jump the legal hurdles that have become the prerequisites of legacy. Had Dolbear been more observant of Patent Office formalities, the editors noted, "it is probable that the speaking telephone, now so widely credited o Mr. Bell, would have been garnered among [Dolbear's] own laurels."[63]

Antonio Meucci (1871)

In a 1999 episode of HBO's *The Sopranos*, Tony Soprano claimed that Bell "robbed" Antonio Meucci of the credit for having invented the telephone. The irony of a fictitious mob boss decrying a robbery aside, Tony Soprano is not alone in his ire, especially among Italian-Americans, many of whom insist that poverty, poor luck, and anti-Italian bigotry denied Meucci his proper place in history.[*]

Antonio Meucci was born into a middle-class Tuscan family in 1808. After studying chemical and mechanical engineering for two years at the Florence Academy of Fine Arts, he could no longer afford the tuition and had to take a job as a stage technician at the Teatro della Pergola in Florence.[64] There, the 17-year-old inventor constructed a type of acoustic pipe-telephone to enable communication between the stage and the control room. However, after Meucci was imprisoned on suspicion (later confirmed) that the inventor was involved in the Italian unification movement, he and his wife, Esterre, quickly fled to Cuba.[65] There, while developing a system for using electrical shocks to treat rheumatism (an affliction his wife suffered), Meucci is said to have invented a device he called a "talking telegraph" that could transmit the human voice through electrical impulses.

In 1850, Meucci and his wife immigrated to the United States, settling in the Italian enclave of Staten Island. Shortly thereafter, Esterre's rheumatism rendered her almost permanently bedridden.

Meucci, who had studied the prevailing literature on electromagnetism and acoustics, soon developed a telephone-like device called a "telettrofono" that would allow his invalid wife to call his basement laboratory from her bed.[66] In his 1857 notes, the inventor described the device as a vibrating diaphragm that altered the current of an electromagnet, which allowed the human voice to be transmitted along a spiral wire to a receiving diaphragm that would reproduce the sound.[67]

In 1858, the Italian painter Nestore Corradi (who Meucci had met when Corradi worked as a set designer in Florence) made sketches of Meucci's device that might help him solicit Italian investors. When no one bit, apparently Meucci published the inventions in the New York City Italian-language newspaper *L'Eco d'Italia* (though no copies of the article have ever been

Antonio Meucci

*   In 2002, former Staten Island congressman Vito Fossello helped pass a congressional resolution (HRes. 269) stating that Meucci's "work in the invention of the telephone should be acknowledged." Although the resolution stopped short of disclaiming Bell's priority, its preamble noted that no patent would have been issued to Bell "if Meucci had been able to pay the $10 fee to maintain the [patent] caveat after 1874." The Senate refused to bring the resolution to a vote, however.

located). Sadly, Meucci was severely injured when a boiler exploded on the Staten Island Ferry. To raise money for his recovery, Esterre sold both the device and Corradi's drawings of it.

Eventually, Meucci raised $20 from several prominent Italian-American entrepreneurs, with whom he partnered in launching The Telettrofono Company. On December 28, 1871, the inventor dispatched his patent lawyer (along with a $15 application fee) to the U.S. Patent Office to file a patent caveat—a kind of legal place-holder until a full patent claim is filed—on a device Meucci labeled the "Sound Telegraph." Although the caveat did not have the level of detail one would expect of a full patent application, it did describe "the employment of a sound conductor, which is also an electrical conductor, as a means of communication by sound between distant points."[68] The caveat, however, was only good for three years, and when it expired on December 28, 1874, Meucci (who was already in great debt) could not afford the $10 renewal fee.[†]

Around 1885, Dr. Seth Beckwith (founder of the Cleveland Homeopathic Hospital) incorporated the Globe Telephone Company with the intention of manufacturing and selling telephonic and electric instruments. It was not long, however, before the Bell Telephone Company took Globe to court for infringing on patents Bell had secured in 1876.[69] In its defense, the company claimed that Bell's patent was invalid because Meucci had invented the phone first.

Unfortunately, the case (and Meucci's claim), got unwittingly wrapped up in a shocking case of public corruption. While serving as Arkansas's senator from 1877 until 1885, Democrat Augustus A. Garland had accepted shares of near-worthless stock in the Pan-Electric Telephone Company, one of several early Bell competitors seeking to curry favor among Washington's most powerful politicians. In 1885, President Grover Cleveland tapped Garland to be attorney general of the United States. It wasn't long before agents of Pan-Electric came knocking at Garland's door requesting that the government official sanction a lawsuit against American Bell Telephone. If Bell's patent was invalidated, Pan-Electric's stock would rise precipitously and Garland was positioned to make a handsome profit. Realizing the potential conflict of interest, Garland declined the request and left D.C. on an extended vacation. While he was away, Garland's deputy, Solicitor General George A. Jenks (apparently unaware of Garland's financial interest), agreed to launch a suit on behalf of Pan-Electric, claiming that Bell's patent was based on fraud.

When word of Garland's ties to Pan-Electric became public, the resulting scandal complicated and delayed not only Pan-Electric's suit, but all those (including Globe Telephone's) that sought to overturn Bell's patent.[70] Although Cleveland won the popular vote in his bid for reelection in 1888, he lost the Electoral College, and Garland lost his position as Attorney General. The new administration eventually dropped the Pan-Electric case, leaving undecided Muecci's claim to the telephone.

Elisha Gray (1876)

A Quaker professor of science at Oberlin College (though he never completed his bachelor's degree), Elisha Gray is best known as the father of the modern musical synthesizer.[71] Among some circles, however, he is known as the man from whom Alexander Graham Bell stole the design for the telephone.

In 1869, Gray (along with his partner, Enos M. Barton) founded a company that supplied telephone

---

Some critics dispute this claim, however, since Meucci somehow managed to scrape together $150 to file full patent applications on inventions unrelated to the telephone between 1872 and 1876. See Bob Estreich, *Antonio*

company's top engineer until 1876, when he decided to focus exclusively on his own inventions. At the time, he was chiefly financed by Dr. Samuel White, a wealthy Philadelphia dentist who hoped to profit off of Gray's improvements on the acoustic telegraph.

Knowing that White wanted him to focus exclusively on improving the telegraph, Gray worked in secret, designing a telephone that used variable resistance to transmit signals. He mentioned his experiments to no one until the morning of Friday, February 11, 1876, when he instructed his patent lawyer to draft a caveat on an apparatus and method for "transmitting vocal sounds telegraphically" using liquid as a kind of transmitter.[72] By Monday morning (February 14, 1876), the caveat was written, signed by Gray, notarized and filed with the U.S. Patent Office. Although there is some dispute about the timing, the consensus is that Bell's lawyer filed a patent application almost identical to Gray's caveat a few hours later.‡ However, the lawyer requested that the processing clerk record and immediately hand-deliver the application to the patent examiner to make it appear that Bell had filed first.[73]

The sharp patent examiner, however, noticed the similar claims and suspended review of Bell's application for 90 days to give Gray time to submit a full application of his own. However, the ploy stilled paid off. During the delay, Bell was summoned to Washington, D.C., by his lawyer. After the two conferred, they paid a visit to the patent examiner, Zenas Fisk Wilber, ostensibly to get some clarification about the competing claims. Although Wilber had officially suspended his review, when confronted, he told Bell about the liquid transmitter idea in Gray's competing caveat and

requested proof that Bell had thought of it first.[74] Bell cited an application filed a year previously in which he suggested the use of mercury in a circuit breaker. Although mercury would not have worked as a telephone transmitter, Wilber apparently accepted Bell's oblique reference to mercury in a completely unrelated patent as evidence of the inventor's priority and approved Bell's application on March 3, 1876, seventy-three days before Gray's deadline expired. Ten years later, in a sworn affidavit, Wilber admitted that he was deeply in debt to Bell's lawyer (whom he had served with in the Civil War), and had allowed him and Bell to view Gray's caveat prior to filing Bell's application.[75]

Elisha Gray

‡   Historian A. Edward Evenson claims that, over the weekend, Bell's attorney somehow learned of the liquid transmitter in the caveat being prepared by Gray's lawyer. Without consulting Bell (who was in Boston at the time and did not even know that the application was being filed), his attorney added a description of a liquid mercury transmitter to Bell's patent application. See Evenson, Edward A., *The Telephone Patent Conspiracy of 1876: The Elisha Gray–Alexander Bell Controversy* (Jefferson, NC: McFarland, 2000): 80–82.

Regardless of this dating error, however, Tesla was neither the first inventor to grasp wireless transmission of electricity, nor even the first to patent the innovation. The hidden history of wireless transmission starts decades before Tesla's Colorado experiments and is wrapped up in the sordid tale of the invention of the telephone.

Wireless Wizardry

Sometime around 1860, a Washington, D.C., dentist (and amateur electrician) named Mahlon Loomis was experimenting with electrical currents in the upper atmosphere, which he thought might be tapped by kites carrying metal wires.[76] He soon discovered, however, that changes in the current of a wire held aloft in one kite appeared to disturb the current of a wire held by a kite some distance away. Loomis thought he could make use of these disturbances as a practical means for long-distance wireless telegraphy.[77] By the end of the Civil War, Loomis had worked out just such a system and, in either 1868 or 1872 (depending on the source), demonstrated it to several members of Congress and prominent D.C.-based scientists by transmitting communication signals between kites separated by about fourteen miles (though there is some discrepancy, as well, regarding the actual distance).[78]

Spurred by the success of the public grant to Samuel S. Morse (of Morse-code fame), the following year Senator Charles Sumner of Massachusetts introduced a bill in the U.S. House of Representatives to incorporate and fund the Loomis Aerial Telegraph Company, with the aim of making Loomis's wireless system practical. Many congressmen felt the whole thing was a scam, however, and stripped any government appropriations out of the bill.[79]

While official incorporation of the Loomis Aerial Telegraph Company was mired in committee, Loomis filed for (and received) a patent on a method for "improvement of telegraphing" which used "the earth as one-half the circuit and the continuous electrical element far above the earth's surface for the other part of the circuit."[80] Because "an interruption of the continuity of one of the conductors from the electrical body" is "indicated upon its opposite or corresponding terminus," Loomis supposed he could produce a communicating circuit between the two terminals without the use of connecting wires or cables.[81]

Loomis's patent application bore an alarming resemblance to one filed three months earlier by the inventor William Henry Ward, who may have heard about Loomis's widely publicized demonstrations before D.C.'s elite. Where Loomis relied on mountaintop aerials or high-altitude kites, Ward suggested using electrical towers.[82] In some places, however, the language used in each application is almost identical. Ward described using his towers to collect electricity "for the use of land lines of telegraphs or *for other purposes, such as light, heat &c.* By use of aerial electricity, I entirely dispense with artificial batteries."[83] Similarly, Loomis wrote, "I also dispense with artificial batteries, but use the free electricity of the atmosphere, cooperating with that of the earth, to supply electrical dynamic force or current for telegraphing and *for other purposes, such as light, heat, and motive power.*"[84] In any case, it was clear that, while both inventors refer to a method for improving telegraph systems, they both envisioned that the same wireless system could be used for transmitting useful electricity between distant points using the atmosphere as a conductor.

It was in the quest for a wireless telephone that the wireless transmission of electricity was inadvertently discovered. In 1882, just before he was sued by American Bell for infringing on Bell's telephone patent (and some ten years after Ward and Loomis filed their patents), Amos Dolbear demonstrated that he could transmit a wireless telegraph signal over a distance of a quarter of a mile using phones connected to metal rods rammed into the earth.[85] By connecting one end of an induction coil to a capacitor and the other end to the ground, Carlson noted that Dolbear was able to transmit telephone signals from his laboratory at Tufts University to his home nearby.[86]

On March 24, 1863, the young professor filed a patent for a "mode of electric

San Francisco Call feature on Tesla's Wireless Transmission

RADAR refers to Radio Detecting And Ranging. It is a system to determine the range, direction and acceleration of an object by interpreting electromagnetic waves that are reflected off of it. Usually radio waves (or sometimes microwaves) are used since they are not as easily absorbed or scattered by air or water as, say, infrared or ultraviolet radiation. Instead, radio waves remain relatively uniform until they are scattered by the reflective surface of the object being detected.

In a June 1900 article for *Century* magazine, Tesla first suggested that stationary electromagnetic waves could be used to "determine the relative position or course of a moving object," an idea that would prove remarkably similar to modern RADAR.[90] However, it was not until August 1917, in an article for *The Electrical Experimenter*, that Tesla first articulated a method for detecting submarines that involved bouncing a concentrated ray of high-frequency electrical charge off of a metallic hull and intercepting the reflection on a fluorescent screen.[91] Consequently, many claim that Tesla is RADAR's true inventor. While acknowledging its development around the globe, biographer Margaret Cheney nevertheless referred to RADAR as "an international achievement inspired by the mind of Tesla."[92]

As early as 1886, Heinrich Hertz had demonstrated that solid objects would reflect radio waves. Then, while performing experiments in radio transmission between two ships in 1897, Alexander Popov, a physicist for the Russian Navy, discovered that he could detect the interference from a third ship passing between the two. At the time, Popov noted that he might be able to use the phenomenon to detect objects from afar, though (like Tesla) he never pursued the idea.[93] In 1904, however, the German inventor Christian Hulsmeyer gave public demonstrations in Germany and the Netherlands of a method to detect radio echoes

from a device that used a dipole antenna to transmit a signal, and a parabolic dish to capture its reflection. He intended that the device be used by the German navy to prevent ship collisions, but failed to gain the interest of any naval authorities.[94] Still, Hulsmeyer received patents for his device in Germany and England that same year, some thirteen years before Tesla provided a cogent explanation for the phenomenon.[95] Recognizing RADAR's usefulness in warfighting, as World War II ramped up, scientists and inventors in Germany, France, Italy, the Netherlands, the Soviet Union, the United States, Great Britain and Japan were independently developing various systems that would evolve into modern RADAR.[96]

Although Tesla spoke only of stationary electromagnetic waves, without defining which frequencies would be appropriate, the use of microwaves for object-detection was spearheaded by an enterprising Japanese physics student named Shintaro Uda. In 1926,

Christian Hulsmeyer

Uda, a graduate assistant to Professor Hidetsugu Yagi at Japan's Tohoku Imperial University, first patented the design for the microwave antenna that became the basis for many designs still used today.[97] Like most graduate assistants, he teamed with his mentor on a paper that outlined the invention and was published by the Imperial Academy of Japan.[98] However, it was Yagi who was invited to speak publicly about the experiments he and Uda had carried out. As a result, the invention came to be known as the "Yagi antenna" and Uda's contribution was (almost) entirely obscured.

Interestingly, although the Yagi (Uda) antenna was widely used for airborne RADAR by allied forces during World War II, internal rivalries in the Imperial Navy resulted in the invention being shelved in Japan. In fact, most Japanese military authorities were not even aware of the design until after the Battle of Singapore in 1942, when a captured British RADAR technician (whose notes mentioned a "Yagi antenna") revealed under interrogation the invention's connection to the Tohoku University professor.[99]

Shintaro Uda

communication" that worked by creating a large voltage difference between grounded terminals in order to send a current "between two or more places without the use of a wire or other like conductor."[87] Although Dolbear noted that his patent spoke primarily of telephone instruments, "as these give the best results," he nevertheless observed that, "any electric instruments may be used capable of utilizing the currents passing through the earth."[88] Thus, not only was the American aware that his method was capable of transmitting useful amounts of electricity wirelessly, he had—some fourteen years before Tesla completed his application for a method of wireless transmission—received a patent on the process. In fact, Carlson speculated that Tesla may have relied upon Dolbear's patent when configuring the wireless set-up in his Fifth Avenue laboratory.[89] In any event, by all legal standards, Dolbear's 1882 patent application anticipated virtually the same method for wireless transmission of electricity as Tesla's later patents. Even if Dolbear failed to make the same fantastic claims that would forever link Tesla with the discovery of wireless transmission, the historical evidence is clear that Dolbear not only thought of the process prior to Tesla, but clearly was the first to patent it.

"The progressive development of man . . . is the most important product of his creative brain. Its ultimate purpose is the complete mastery of mind over the material world, the harnessing of the forces of nature to human needs. This is the difficult task of the inventor, who is often misunderstood and unrewarded."

—NIKOLA TESLA, *MY INVENTIONS*, 1919

chapter seven

The Truth about Tesla

On December 17, 1903, while Tesla was desperately trying to stay one step ahead of his creditors, Orville and Wilbur Wright made the first controlled, powered, and sustained heavier-than-air human flight. All of those limiting adjectives are necessary because the Wright brothers certainly were not the first to invent, or even successfully fly, an experimental manned aircraft.[1] Sometime prior to 1849, English engineer Sir George Cayley demonstrated heavier-than-air manned flight by sending a ten-year-old boy aloft on a triplane glider.[*] In August 1901 (more than two years before the Wright brothers made their historic flight), six people—including a reporter for the *Bridgeport Herald*—claim to have witnessed German inventor Gustave Whitehead (Weisskopf) in Connecticut flying his Condor No. 21 "flying car" fifty feet off the ground for a mile and a half.[2]

While others may have been first in the air, the Wright brothers made one crucial innovation: "wing warping," a method of maintaining lateral stability using a series of pulleys to twist and warp the wingtips. This warping allowed an airplane to make banked turns and stabilize itself from gusts of wind, and it represented a groundbreaking step in aviation. But, in March 1903 (more than nine months before their flight), the brothers applied for a patent not only on this *means* of lateral control, but on the *idea* of lateral control itself. So broadly was the patent claim written that it covered virtually any mechanism an inventor could ever possibly devise for lateral stability in an airplane.[†]

---

[*]   Four years later, the seventy-nine-year-old Cayley flew either his grandson or his young coachman—depending on whose report you read—across Brompton Dale for about 180 meters before crashing.

[†]   Indeed, the first—of eighteen total—claims the Wrights' application covered was for "a normally flat aeroplane having lateral marginal portions capable of movement to different positions above or below the normal plane of the body of the aeroplane, such movement being about an axis transverse to the line of flight, whereby said lateral marginal portions may be moved to different angles of incidence, and means for so moving said lateral marginal portions . . ." See Orville and Wilbur Wright, Flying Machine, U.S. Patent No. 821,393, filed March 23, 1903, and issued May 22, 1906.

Between 1903 and when the Wright brothers were finally granted their patent in 1906, several inventors worked on innovations of their own, some vastly superior to anything the Wrights ever concocted. Glenn H. Curtiss, a motorcycle engineer who began designing dirigible engines, for example, thought to add takeoff and landing gear to an airplane's bottom, and invented an entirely different method of lateral stability. By raising or lowering the flaps on the wings, a pilot could control the roll of an airplane and achieve far sharper banking turns than the Wright brothers' wing-warping method. (Wing flaps proved so effective, in fact, that they have become the lateral control method of choice for modern airplane manufacturers.)[3]

The broad patent issued to the Wright brothers did little to stop Curtiss, who continued to design and manufacture airplanes to sell to the Aeronautic Society of New York until 1909, when the brothers brought a patent infringement suit in the federal district court for Western New York. The case was decided by Judge John R. Hazel, the same judge who overturned the Second Circuit Court of Appeals' decision to uphold Tesla's patent on his split-phase AC motor design.[4] Keeping to the letter of the Wright brothers' patent, Hazel introduced a new and powerful concept in American jurisprudence: that a competitor could infringe on a patent not only by employing the method explicitly claimed in the patent, but by employing any means that were "equivalent" to the method.[5]

The patent case (and Judge Hazel's subsequent decision) brought aviation innovation to a screeching halt, at least in the United States. In Europe, where patent enforcement was lax, improvement on airplane design continued unabated, leading the Germans (in particular) to develop and manufacture vastly superior airplanes. It took World War I, and the intervention of the U.S. military in the name of international security, to wade through the mire and help the U.S. (and, subsequently, the Allied Powers) to catch up. They created

Gustav Whitehead with his Condor No. 21 Flying Car, 1901

a patent pool whereby every inventor could improve on existing designs and every manufacturer was guaranteed a portion of the profits. Even so, Judge Hazel's legal precedent had set American aviation back so far that many historians believe that the Germans (despite military defeat and a crippling economy) were still ahead by the time hostilities broke out again in 1939. Nevertheless, to this day, the Wright brothers are credited with having invented flight and heralded with ushering in the modern miracle of manned aviation.

Mythbusting

Like the conventional account of the Wright brothers, the conventional mythology of Nikola Tesla is a kind of shorthand for a historical record as complicated as the man himself. It is a shorthand that relies largely on the outcome of complex patent disputes without acknowledging or appreciating either the scientific nuances that are the basis of the lawsuits, or the social and political forces that contribute to their resolution.

If the history of innovation is not written *by* the winners, it is certainly written *about* them. For a long time this was also true of Tesla, who, by the end of his life had lost many of his friends, most of his fame, and all of his fortune. Once lauded as a scientific virtuoso on par with Isaac Newton, Tesla is said to have been all but erased from the annals of history—at least in the United States. In his extensive biography, Carlson attributed the inventor's relative obscurity in the second half of the twentieth century variously to capitalism (he never had a successful company), xenophobia (he was not born in America), and

populism (a recluse and a dandy, he was often seen as eccentric and effete).[6]

Still, shortly after Tesla's death, Pulitzer Prize–winning author John O'Neill wrote *Prodigal Genius: The Life of Nikola Tesla*, the first comprehensive biography of the inventor. By all accounts it was a popular and commercial success. In 1959 and 1961, two short biographies of Tesla meant for young readers were published. A more exhaustive record, with better documentation and photographs from Tesla's time in Colorado, was penned by Inez Hunt and Wanetta Draper in 1964.[7] Although a handful of foreign-language biographies were published after 1964, none was released in the United States until Margaret Cheney's seminal account *Tesla: Man Out of Time* in 1989. Since then, no fewer than fourteen biographies of Tesla (in English) have come out, including Marc Seifer's exhaustive *Wizard: The Life and Times of Nikola Tesla, Biography of a Genius* and W. Bernard Carlson's recent *Tesla: Inventor of the Electric Age*.[‡]

According to Carlson, Tesla's reemergence in popular culture (if he was ever really gone[§]), ironically may be due to his outsider status.[8] At least since the energy crisis of the early 1970s, he has been embraced by the American counterculture in large part because he was rejected by "establishment figures" like Thomas Edison and J. P. Morgan. For his part, Carlson attributed Tesla's initial urge to con-

---

‡   While Tesla biographies have increasingly noted the inventor's personal and professional failures, they are universally commendatory.

§   Even by 1989, Margaret Cheney noted how "astounding" it was that a new generation of technocrats had pounced on Tesla's life and inventions, venerating him as a genius ahead of his time: "Today one can hardly pick up a copy of *The Wall Street Journal* or *The New York Times* without finding a mention of [Tesla] or his effect on famous young followers and admirers." See Cheney, *Tesla*, xi.

coct groundbreaking innovations to his desire to achieve "insider status" among the respected New York society of electrical engineers.[9] (Once Tesla attained this status, however, Carlson speculated that his inventive instinct was driven more by a deep-seated psychology: a desire "to reorder the world as a means of compensating for the disorder he felt inside.")[10]

Whatever his creative motivations, Tesla is roundly regarded as an inventor of singular genius, independently conceiving radical, new technologies "from within," despite all evidence that many of his ideas were not new, and that even his novelties built upon concepts that far predated him and were the product of an evolutionary process involving many minds.[11] Carlson, for example, concluded that Tesla's AC motor allowed utilities to shift from direct current to alternating current and expand their services from electrical lighting to electrical power for all sorts of industrial and consumer uses.[12] This is a rather bizarre claim, given that Carlson acknowledged that there were at least fifteen separate firms manufacturing more than ten thousand electric motors by 1887, well before Tesla had patented his own motor design.[13] Clearly, before Tesla revealed his groundbreaking innovation, momentum was already behind a transition to industrial and consumer electricity. Moreover, the proponents of AC were winning the War of the Currents even before Tesla won his first patent on the polyphase AC motor in May 1888. Even so, Carlson documented how Tesla's initial design, which required at least two independent AC generators and double the copper wiring of a DC system, was regarded as commercially impractical by Tesla's business partners (as well as by every manufacturer they approached). Thus, Tesla's contribution to the transition to industrial and consumer power may be far less consequential than some historians suggest.

Carlson counted as equally innovative Tesla's introduction of "the idea" of polyphase power, which allowed for the efficient transmission of AC power over long distances.[14] But as early as the Turin Electrical Exhibition of 1884, Lucien Gaulard and John Gibbs had demonstrated their transformer design on a single-phase transmission circuit some forty-eight miles long.[15] A year later, Galileo Ferraris (who had organized the Turin expo) conceived of using multiple, out-of-phase currents to produce a rotating magnetic field. A year after that, the wily engineers at the Ganz Works installed the first long-distance single-phase AC transmission system, using a ZBD transformer to dispatch high-voltage electricity from Cerchi to Rome, some seventeen miles away.[16] Polyphase power may have made AC transmission even more efficient, but it is clear that, even before Tesla introduced his split-phase design in December 1888, most of the engineering community already had conceded AC power's superiority.

If Tesla's key insight was transmitting power in multiple phases, it is certainly unclear that he was first to devise the idea or that he did so independently. Indeed, even if we were to believe Tesla's own account that he had a Eureka moment while walking through a Budapest park in 1882, the first time he mentioned a multi-phase AC motor design was six years later, in April 1888, during a conversation with his patent lawyer James Page. For all intents and purposes, no one was interested in manufacturing Tesla's original design prior to this revelation.

Justice Wiley Blount Rutledge

At the exact time Tesla claims to have had this conversation with Page (which Page could not convincingly corroborate when pointedly questioned), Galileo Ferraris was presenting his groundbreaking paper on polyphase transmission to the Royal Academy of Sciences in Turin. It was not until almost eight months after Ferraris's speech that Tesla applied for a patent on the design that Carlson claimed would "set the stage for the ways in which we produce and consume electricity today."[17] Whatever role he might have played in the drama of AC power transmission, it is clear that the stage was set not by Tesla, but by several European inventors who remain as obscure as Tesla ever was.

Invention and Innovation

In 1942, President Franklin Delano Roosevelt appointed Wiley Blount Rutledge, dean of the University of Iowa College of Law, to the U.S. Supreme Court. Justice Rutledge had barely settled into his chambers when he was called upon to render an opinion in the *Marconi Wireless* case. Joining Justice Frankfurter, Rutledge wrote an opinion dissenting, in part, from the majority holding that Tesla's radio patents precluded Marconi's. Like Frankfurter, Rutledge cautioned against examining technological innovations with twenty-twenty hindsight. He preferred, instead, to rely upon the "careful, considered and contemporaneous judgment" of the professional clerks who first granted Marconi's patent.[18] Rutledge's decision in favor of Marconi rested on a dogged pragmatism. Marconi was the first to make wireless telegraphy commercially feasible. His device was, quite simply, the first to experience immediate and vast success.

Still, in his meticulous dissent, Rutledge noted that, while Marconi benefited from placing the single straw that broke the camel's back, he owed his success to the many engineers and inventors—Tesla included—who had burdened the camel with bales of their own:

> *There was not room for any great leap of thought, beyond what [Marconi] and others had done, to bring to birth the practical and useful result . . . The invention was, so to speak, hovering in the general climate of science, momentarily awaiting birth. But just the right releasing touch had not been found. Marconi added it.*[19]

For that meager contribution, Rutledge acknowledged, our system of patent law demanded that Marconi's patent claim be upheld.

Rutledge may not have known it at the time, but he had unwittingly stumbled upon the conflation of invention and innovation, a confusion that plagues our understanding of technological progress even today. As Thomas Kuczmarski, a former professor of innovation at Northwestern University's Kellogg School of Management, has noted, "invention and innovation have been mashed together so thoroughly that it is hard to tell the difference between them."[20] In economic terms, invention generates new ideas, designs, or things. It is pure creation. Innovation, on the other hand, adds value to ideas, designs, or things that are already invented. Alexander Graham Bell may have "invented" the telephone (though the jury may have been hoodwinked on that one), but he assumed that people would use it mostly for listening to live concerts from afar.[21] When, instead, people used it to talk to each other, they were engaging in the process of innovation.

The difference between invention and innovation is a subtle one, and one which American patent law often confuses. Indeed, it may have been the very foundation for the legal flip-flop-flip over Tesla's patent on the polyphase AC motor. It was certainly enough to split the Supreme Court five-to-three on just who should get credit for radio. The process of invention, like innovation, is more often than not incremental, building upon existing knowledge. Rarely is either the result of discrete ideas conceived in isolation.[22] It is not surprising, then, that inventors are likely to have similar insights almost simultaneously. Great minds think alike precisely because they are relying on the same body of work that came before them.[23]

American patent law, preoccupied as it is with promoting innovation without stifling invention, is a clunky tool for assessing the rather intuitive process that drives both. It pretends that detached flashes of brilliance can separate invention from innovation and struggles to maintain this convenient façade. Generally, though, the two cannot be so easily divorced. That is why the confusion of patent law makes for a poor gauge of historical significance. The history of innovation comprises everything that happened *prior* to a patent case, not only what determined its resolution.

When Tesla patented his "oscillating coil" and wireless "system of transmission of electrical energy," was he engaged in invention or innovation? The answer is neither—and both—an inconvenient ambiguity that smashes to smithereens the edifice of U.S. patent law. Ironically, the best evidence of the incremental process by which Tesla invented (and innovated) both his induction coil and his system of wireless transmission may be found in the evolution of his patent applications. It is clear that the ideas behind electrical oscillation, resonance, and electromagnetic radiation were "hovering in the general climate" after James Maxwell published his groundbreaking paper in 1865—and certainly after Heinrich Hertz's experiments proved Maxwell's theories in 1891. Tesla's first patent on what would become his oscillating transformer merely added a large spark gap between the windings, a relatively simple, if impractical, modification of an induction coil design that had been around since the 1840s. This innovation was obvious enough that it essentially mimicked equipment Hertz had designed to detect electromagnetic waves in 1887. Although it was not until 1891 that the German would be credited with proving Maxwell's theories, he was sending papers to Helmholtz for publication as early as November 1887. There is no doubt that Tesla was aware of Hertz's work, since we know that he repeated Hertz's experiments sometime in 1890.[24]

Hertz's equipment worked, in part, because of his understanding of resonant frequencies, a concept that would not find its way into Tesla's induction coil patents until 1893. Even then, Tesla incorrectly attributed the phenomenon to waves of energy that propagated by conduction. The inventor went to his death believing that electromagnetic fields required an undetected "ether" to travel, and he roundly rejected quantum mechanics.[25]

It was Tesla's misunderstanding of electromagnetic radiation that first inspired him to look into ground currents as a method of wireless transmission, a concept that free-energy fanatics attribute almost exclusively to Tesla. But the practical application of ground currents for transmitting electrical impulses had been proven by Carl von Steinheil as early as 1836, over sixty years before Tesla filed his first patent on a wireless transmission system. Even then, Tesla did not fully understand (or appreciate) standing waves, a concept critical to resonant frequencies.

Many historians date Tesla's "invention" of wireless transmission to his 1897 patent (filed before he left New York for Colorado), overlooking that the critical components for achieving electrical resonance were not added until he amended his filing in 1900 (after he

had returned from Colorado). Tesla's original apparatus was a slight variation on designs demonstrated by William Henry Ward and Mahlon Loomis forty years before, and almost identical to the wireless telephone Amos Dolbear patented in 1863. Dolbear may have viewed his wireless design as an innovation on the telephone, but he was not oblivious to the fact that it could—given high enough voltages—be capable of transmitting electrical power.

Tesla thought larger than Dolbear, that is for certain. His idea of a worldwide system for transmitting electricity (and communications) via resonant waves in the earth was big, bold, even romantic. The conventional myth is that J. P. Morgan and other clandestine financiers conspired to destroy Tesla's vision because it would have been impossible to meter the resulting energy. This mythology leads some Tesla enthusiasts to espouse the inventor's "World-System" as a sort of free-energy panacea that threatened the greedy capitalist structure.[26]

The truth is the inventor's wireless power system most likely would have failed, even if Morgan had fully funded Tesla's vision. Tesla incorrectly assumed that resonant waves in the earth would not dissipate, allowing efficient transmission of energy from the point of generation to any place on the earth's surface. Carlson analogized Tesla's hypothesis to a water balloon. Tesla assumed that the earth behaved like it was filled with an incompressible fluid.[27] Push water into a hole on one side and it would come squirting out a hole on the other. Instead, the earth behaves like it is filled with snow, which can be compressed into a denser and denser ice ball. Thus, pumping electrical waves into the earth (at least as Tesla envisioned it) would not have overcome the same dissipation problem that plagued William C. Brown and his team when they set about trying to design a solar-powered satellite for Werner von Braun in 1967 (and, incidentally, the many scientists that have chased point-to-point wireless power transmission ever since).

The truth is that, while Tesla certainly was subject to many unfortunate circumstances (like the Navy failing to pay for the licenses of his radio designs), he is largely responsible for his own failures. His first—and, perhaps, greatest—financial mistake was agreeing to waive the $2.50-per-horsepower royalty on each motor that he had negotiated with Westinghouse. Understandably, Westinghouse was in a dire fiscal situation of his own by 1890 and may have gone bankrupt trying to pay the fees. But, rather than waive them altogether, Tesla certainly was savvy enough to have negotiated lower (or delayed) royalty payments. Instead, he trusted that Westinghouse would bankroll all of his future endeavors after the industrialist reaped the windfall of Tesla's inventions. Of course, Westinghouse's largesse never really materialized.

Tesla could have lived more frugally. He could have better employed Astor's investment by staying in New York City and commercializing the innovations he had developed in AC lighting rather than taking Astor's money and running off to Colorado to perform experiments that, ultimately, provided little guidance in developing a wireless power system. His decision to escape to Colorado not only squandered funds he could have invested in his existing patents (or could have used to fight Marconi's radio patent claim), but it

also burned his relationship with Astor and rightfully earned Tesla a reputation as a risky investment for potential financiers.

Tesla's most vocal admirers should keep in mind that the inventor's poor financial decisions were not the unfortunate bumblings of an altruistic genius more concerned with perfecting his ideas than profiting from them. This worn trope is carted out by historians and enthusiasts alike to sustain a caricature of inventors as somehow living above the economic reality in which the rest of us are mired (though, tragically, they remain victim to it). Certainly Tesla was financially savvy enough to withhold (so he claimed) evidence of his split-phase motor design in order to better reap its economic reward. He was fiscally minded enough to plan a European tour designed, in part, to shore up his priority claim on the motor's patent. Carlson went to great lengths to examine how Tesla used spectacle and illusion to sell his innovations, both literally and figuratively.[28] It is difficult to square Tesla's P. T. Barnum-esque showmanship with the stereotype of the idealistic and naïve inventor, an oblivious pawn in the chess game of moneyed interests.

Ironically, Carlson used the strained relationship between Tesla and Morgan (and Adams, and Astor, and Westinghouse) to examine the relationship between all inventors and their investors, suggesting that inventors who cannot fully conceive their ideal turn to illusion to convey a vision that is inspiring, though not entirely complete. "In sharing an ideal with others," Carlson wrote, "inventors have to confront [ambiguity] head-on: if they cannot fully access the ideal, how can they convey it to friends, patrons, patent examiners, and customers?"[29]

Archimedes had an epiphany—the original Eureka moment—while soaking in a bathtub

Law and Legacy

And here we are brought full circle. The process of invention (and innovation) involves ambiguity; it practically drowns in it, like Archimedes in an overflowing bathtub. Invention and innovation require playing with ideas that are neither novel nor practical, flights of fancy through "the general climate of science" in which great ideas hover. But, as Carlson (and Mark Twain) accurately recognized, ambiguity is a square peg in patent law's round hole. Indeed, the great mind hailed as the shoulder upon which human progress has stood (to mix anatomical metaphors) is he who "added his little mite," she who placed the single, back-breaking straw.

History is not patent law, nor should it be. If there is anything that must embrace ambiguity, it is the story of how we got here. It is clear that, just as Justice Frankfurter noted more than a century ago, the great transforming forces of technology that are shaping the twenty-first century have rendered obsolete much of U.S. patent law. Perhaps more importantly, they have called into question the process by which history is written, where legacy is bestowed by legal victories that barely reflect the lived experience of real people. If a more thorough exploration of Tesla's life and achievements has revealed anything it is that technological innovations—from fluorescent lighting to the telephone, from the airplane to radar—are not the singular achievements of uniquely great minds, but the product of a cacophony of thinkers, and our popular history ought to attest to this revelation.

Changing the way we record the history of innovation does not require us to become cynical iconoclasts set on tarnishing the memory of great figures like Nikola Tesla. Rather, it requires us to better appreciate the messy process of invention and innovation, a process that involves inflated egos and flawed personalities. It requires us not to abandon our golden calves—idols of invention that were never more than apparitions of imagination to begin with—but to recognize and revere the multitude of imperfect humans, not unlike ourselves, who, through trial and tribulation, error and achievement, propelled us ever forward in the hard-wrought progression of humanity.

Afterword

Five Oft-Repeated Myths about Nikola Tesla

One of the main theses of this book is that historians have relied upon the outcome of patent disputes to perpetuate the myth of Nikola Tesla as the lone genius inventor, even while overlooking obvious evidence to the contrary. But Tesla's biographers should not carry all the blame for this mythology. In most cases, their publications are far more nuanced, sometimes failing to reach obvious conclusions, but generally containing accurate historical accounts.

The academic rigor demonstrated by Tesla historians is often lacking in Tesla enthusiasts. While researching this book, I joined a number of online forums where these enthusiasts tend to venerate the inventor while discussing various utopian "free" energy fantasies. Rather quickly, I was faced with the realization that the vast majority of these well-meaning individuals apparently had read very little about Nikola Tesla.

OPPOSITE: Tesla monument in Queen Victoria Park, Ontario, Canada

Most were simply repeating conspiracy theories and well-worn tropes, most often involving Thomas Edison, J. P. Morgan, or some faceless group of evil financiers bent on thwarting Tesla's inventions.

Invariably, these forums include two or three electrical engineers, or individuals who have read enough to be considered Tesla scholars, of sorts. These exhausted few spend an inordinate amount of time explaining Electricity 101 and correcting ad nauseam the most repeated inaccuracies masquerading as common knowledge. It is for these individuals that I gather here the most salient answers to five myths about Nikola Tesla that I have read most often in these forums.

MYTH #1:
Nikola Tesla "Invented" Alternating Current (AC)

Perhaps no single myth is more often repeated than the claim that Nikola Tesla invented alternating current. Pull anyone from off the street and this inaccuracy is what they know about Tesla (if they know anything about the inventor at all). Certainly there is good reason for Tesla's name to be forever associated with alternating current. The infamous "War of the Currents" makes for good storytelling, pitting the cleverest inventors of the late nineteenth century against one another in an epic struggle over the future of electricity. But the truth is far less dramatic.

• The first electromagnetic generators were invented almost thirty years before Tesla was born and, because they universally relied on rotary mechanical motion (like a hand-crank), all of them generated alternating current. Because almost all electrical devices had been developed to run on electrochemical batteries, however, no one had any use for alternating current. In fact, by the time Tesla even began learning about electricity, all electromagnetic generators were fitted with a device called a commutator used to convert alternating current to more useful direct current (DC).

• Before Tesla ever patented his AC-powered motor, most electrical systems were used strictly for lighting purposes and ran on DC. Nevertheless, by the time Tesla was working on his motor, AC power was already in widespread use. George Westinghouse's company was selling AC systems that utilized transformers designed by Lucien Gaulard and John Gibbs more than five years before Tesla patented his system. In fact, before Tesla completed the design on his system, Westinghouse claimed to have "sold more central [power] stations…on the alternating current system than all of the other electrical companies in the country put together on the direct current system."

MYTH #2:
Thomas Edison Thwarted Tesla to Protect His Interests in Direct Current (DC)

The short-hand version of the War of the Currents has a greedy, manipulative Thomas Edison squaring off against a meek and benevolent Nikola Tesla. Edison had sunk his investments into the less efficient direct current (DC) and used lies and evil tricks to undermine Tesla and his alternating current system. While Tesla's AC system eventually won the day, Edison's efforts ensured that Tesla never profited from the invention. This Manichean tale of good and evil may simplify history, but the truth is far less black and white.

• The War of the Currents was well underway long before Tesla ever patented his AC system. By 1885, George Westinghouse was feverishly selling an AC power system designed in part by the brilliant engineer William Stanley and utilizing transformers designed by Lucien Gaulard and John Gibbs.

• Even before Tesla's system was available, Edison and Westinghouse faced stiff competition from Elihu Thomson, who, together with his high school colleague Edwin J. Houston, had formed the Thomas-Houston Electric Company and were successfully manufacturing and selling their own commercial AC power system.

• Most of Edison's more questionable attempts to discredit AC power occurred before Tesla had completed the design for his AC power system in 1888. As early as 1886, Edison was publicly decrying the dangers of high-voltage AC power and had hired Harold P. Brown to design the electric chair as a means for humane execution. By

1887, Edison had convinced Elihu Thomson of the dangers of AC and Thomson gave a lecture before the American Institute of Electrical Engineers (AIEE) decrying the public safety risks associated with widespread adoption of AC.

◆ Despite Edison's attempts, AC power became the electricity of choice and Tesla's system was widely adopted. The inventor was set to make millions (perhaps billions of dollars) from royalties on each installed horsepower of AC. In 1891, however, Westinghouse was facing personal financial ruin, and the terms of Tesla's contract threatened to bankrupt the financier. To ensure that his AC system continued to be manufactured, Tesla agreed to waive the royalties entirely. This agreement with Westinghouse was far more responsible for Tesla's impoverishment than anything Edison ever did.

MYTH #3:
Nikola Tesla won the Nobel Prize for his Groundbreaking Inventions

To this day, many people mistakenly claim that Nikola Tesla won the Nobel Prize for his groundbreaking invention of AC power. Some versions of the myth have Tesla turning down the prize after learning he was to share it with his nemesis, Thomas Edison. Another version has Tesla turning down a joint-prize with Guglielmo Marconi for their invention of radio. No matter which story is told, all are wrong. Despite his contributions to electrical innovation, Tesla never won the Nobel Prize.

◆ A November 6, 1915, Reuter's News Agency report from London announced that the 1915 Nobel Prize would be awarded to Thomas Edison and Nikola Tesla for their contributions to the development of consumer electrical systems. The *New York Times* subsequently announced as much in a front page article. However, the following week the Nobel Committee announced that the prize would go to Sir William Henry Bragg and William Lawrence Bragg for their achievements in analyzing crystal structures through X-rays. Though many speculated that the Nobel Committee changed its mind after learning that either Edison or Tesla (or both) would decline the prize, the Nobel Foundation called the rumor "ridiculous" and noted that a recipient had to be awarded the prize before he could decline it.

◆ According to records from the Royal Academy in London, none of the nineteen scientists on the 1915 Nobel physics committee even nominated Tesla for the prize.

◆ Much of the claim that Tesla won the 1915 Nobel Prize can be blamed on careless journalists who repeated the inaccurate claim in subsequent articles. In a December 8, 1915, *New York Times* article promoting Tesla's wireless tower, the author incorrectly described Tesla as "the inventor, winner of the 1915 Nobel Physics Prize."

MYTH #4:
Nikola Tesla Perfected Wireless Transmission of Electricity

Common mythology among Tesla enthusiasts is that the inventor was well on his way to constructing a worldwide system of wireless electricity when shady Wall Street interests realized that the system would cost them millions and so they set about to destroy Tesla's project, as well as his reputation. Despite little understanding of the physics involved with wireless transmission, the conventional wisdom among these individuals is that Tesla's system would have worked, if only it had been completed. In reality, Tesla's wireless vision probably would not have worked, and would likely have been hopelessly expensive and inefficient even if it did.

• Some of the best electrical engineers (and Tesla scholars) on the planet—including Dr. Aleksandar Marincic, Robert Golka, and Leland Anderson—have studied Tesla's magnifying transformer and concluded that the inventor's plan for worldwide wireless transmission was simply impractical. Largely, this is due to the fact that the Earth is not a particularly efficient electrical conductor. Its variable density has a tendency to dissipate electrical waves and disrupt resonant frequencies so that it is virtually impossible to create the kinds of standing waves that Tesla's system required.

• Tesla never successfully tested such a wireless transmission system at Colorado Springs, and the limited experiments he did complete there were likely misinterpreted. Leland Anderson, a senior member of the Institute of Electrical and Electronics Engineers, concluded that the standing waves that Tesla claims to have detected in Colorado Springs were probably transmissions from Tesla's tower rebounding off of Pikes Peak (fortuitously located just across the plain), rather than resonant frequencies rebounding from an antipode halfway around the globe.

• While wireless transmission is certainly possible, it is hopelessly confounded by dissipation. Just like a radio signal weakens as you drive farther away from the transmission tower, electromagnetic radiation dissipates with the square of the distance from

their source. The radius of a sphere of expanding electromagnetic energy, in other words, is the distance from the source of radiation. Since the energy transmitted by any given electromagnetic wave is distributed evenly across an expanding spherical boundary, the amount of energy at any given point is determined by the "inverse square law"—the square of the distance from the power source. Thus, a transmitting tower would have to pump a sufficient amount of energy along the entire sphere, continuously, to ensure that enough energy is available to power a device plugged in anywhere along the radius of the sphere. Such a system wastes far more energy than even the worst transmission wires today.

MYTH #5:
J. P. Morgan Killed Tesla's Plan to Provide the World with Free Energy

The conventional mythology of Nikola Tesla has the inventor squaring off against powerful financial interests, who realized that Tesla's vision of a world-wide system of electricity transmission would be impossible to meter. J. P. Morgan, the primary investor in Tesla's Wardenclyffe Tower, is most often identified as the mastermind behind the conspiracy to destroy Tesla's dream. In fact, Tesla enthusiasts often claim that J. P. Morgan pulled the funding for Wardenclyffe once he realized the implications of Tesla's system. There is little truth in this claim.

• Tesla approached Morgan for funding to develop two wireless towers, one on each side of the Atlantic, which the financier could use to signal steamships at sea and obtain instantaneous stock quotes from the New York Stock Exchange. Initially, Tesla indicated that he could build such a system for as little as $100,000. It was Morgan who balked at the estimate and decided to give Tesla $150,000 to develop a single transmitting tower at Wardenclyffe. However, Morgan was aware of Tesla's tendency to spend money on his own research projects without ever using it for its intended purpose. So he stressed to Tesla that the $150,000 was firm, and he would not be bilked into paying more.

• Tesla immediately set about planning an entire worldwide wireless telegraphy system, including plans for a model city at Wardenclyffe stretching over eighteen hundred acres, with stores, civic buildings and housing for over twenty-five hundred workers. Despite telling Morgan that he could build the tower for $100,000, Tesla's

calculations indicated that a tower capable of worldwide transmission would need to be six hundred feet high, taller than any building then-existing in America. The cost for such a colossal construction ballooned to over $450,000.

• Although Morgan (as he warned) refused to be bilked into paying more than the original estimate, he did agree to allow Tesla to bring on new investors to make up the difference, despite the fact that they would essentially dilute Morgan's interest. After years of making wild claims that never materialized, Tesla could find no one interested in investing in the tower. If Morgan were truly bent on thwarting Tesla's efforts, he would have exercised his majority-control over the enterprise and refused to allow Tesla to solicit more funding.

No. 645,576.

N. TESLA.

Patented Mar. 20, 1900.

SYSTEM OF TRANSMISSION OF ELECTRICAL ENERGY.

(Application filed Sept. 2, 1897.)

(No Model.)

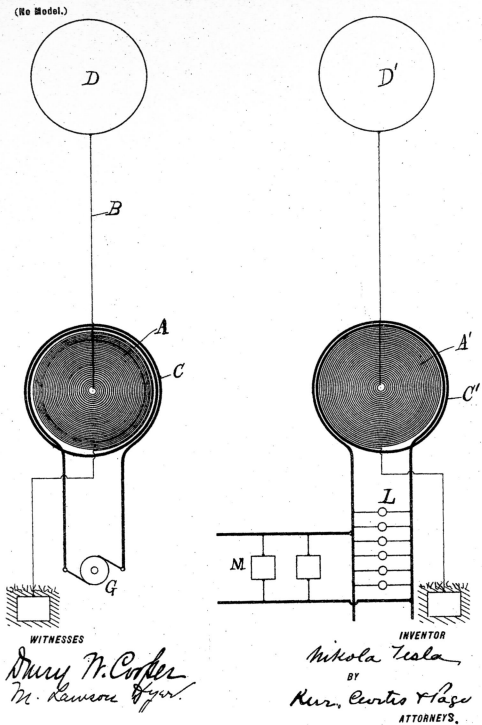

WITNESSES

Drury W. Cooper

M. Lawson Dyer

INVENTOR

Nikola Tesla

BY

Kerr, Curtis & Page

ATTORNEYS.

ENDNOTES

Introduction

1   Edwin Armstrong quoted in Hugo Gernsback, "Nikola Tesla: Father of Wireless, 1856-1943," *Radio Craft*, February, 1943.

2   C.E.L. Brown. Reasons For the Use of the Three-Phase Current in the Lauffen-Frankfurt Transmission, *Electrical World*, 11/7/1891, p. 346.

3   Tesla's Split-Phase Patents, *Electrical Review*, April 26, 1902, p. 291; *Wizard*, pp. 24-26.

4   Leland Anderson (Editor). "John Stone Stone on Nikola Tesla's Priority in Radio and Continuous Wave Radiofrequency Apparatus," *The Antique Wireless Review*, vol. 1, 1986.

Chapter 1

1   Nikola Tesla, System of transmission of electrical energy, U.S. Patent No. 645,576, filed September 2, 1897, and issued March 20, 1900.

2   Margaret Cheney and Robert Uth, *Master of Lightning* (New York: Barnes & Noble Books, 1999), 68.

3   *Ibid*.

4   "Tesla Sues Marconi on Wireless Patent: Alleges That Important Apparatus Infringes Prior Rights Granted to Him," *New York Times*, August 4, 1915, 4.

5   Judge Harlan Stone, (1943). *Marconi Wireless Telegraph Co. of America v. United States*, 320 U.S. 1, 15–16.

6   Judge Felix Frankfurter, *Marconi Wireless Telegraph Co. of America v. United States*, 320 U.S. 1, dissenting in part (1943), 62.

7   Christopher A. Harkins, "Tesla, Marconi, and the Great Radio Controversy: Awarding Patent Damages Without Chilling a Defendant's Incentive to Innovate," *Missouri Law Review 3*, no. 745 (Summer 2008), 760–61.

8   David Kline, "Do Patents Truly Promote Innovation?" *IPWatchdog*, April 15, 2014, http://www.ipwatchdog.com/2014/04/15/do-patents-truly-promote-innovation/id=48768/.

9   *Ibid*.

10   Walter G. Park and Douglas C. Lippoldt, "Technology Transfer and the Economic Implications of the Strengthening of Intellectual Property Rights in Developing Countries," *OECD Trade Policy Working Papers*, no. 62 (2008), http://nw08.american.edu/~wgp/park_lippoldt08.pdf.

11   Carliss Baldwin and Eric von Hippel, "Modeling a Paradigm Shift: From Producer Innovation to User and Open Collaborative Innovation," *Harvard Business School working paper 10-038* (November 2009), 5, http://www.hbs.edu/faculty/Publication%20Files/10-038.pdf.

12   Andrew W. Torrence and Bill Tomlinson, "Patents and the Regress of Useful Arts," *Columbia Science and Technology Law Review* 10, no. 130 (March 15, 2009), http://stlr.org/archived-volumes/volume-x-2008-2009/torrance/.

13 See Dean Keith Simonton, *Scientific Genius: A Psychology of Science* (New York: Cambridge University Press, 1988), http://books.google.com/books?id=cUm4piWluecC&dq=simonton+chance+-configuration+of+genius&source=gbs_navlinks_s.

14 Jeff Dance, "5 Reasons Why Collaboration Contributes to Innovation," *Freshconsulting.com*, September 27, 2008, http://www.freshconsulting.com/5-reasons-why-collaboration-contributes-to-innovation/.

15 Joshua Wolf Shenk, "The End of 'Genius,'" *New York Times*, July 19, 2014, http://www.nytimes.com/2014/07/20/opinion/sunday/the-end-of-genius.html?_r=0.

16 Alfonso Montuori and Ronald E. Purser, "Deconstructing the Lone Genius Myth: Toward a Contextual View of Creativity," *Journal of Humanistic Psychology* 35 (1995): 69, 79, http://jhp.sagepub.com/cgi/content/abstract/35/3/69.

17 *Ibid.*, 74.

18 Mark Lemley, "The Myth of the Sole Inventor," *Michigan Law Review* 110 (2012): 709, 714.

Chapter 2

1 Nikola Tesla, *My Inventions*, ed. David Major (San Bernardino, CA: The Philovox), 2013, 7. First published 1919.

2 W. Bernard Carlson, *Tesla: Inventor of the Electrical Age* (Princeton: Princeton University Press, 2013), 17–18.

3 *Ibid.*, 15.

4 Tesla, *My Inventions, 10–11.*

5 Carlson, *Tesla, 33.*

6 *Ibid.*, 26.

7 Tesla, *My Inventions, 24* (emphasis added).

8 Carlson, *Tesla, 30.*

9 *Ibid.*, 32.

10 Tesla, *My Inventions, 32.*

11 *Ibid.*, 18.

12 Margaret Cheney, *Tesla: Man Out of Time* (New York: Simon & Schuster, 1981), 39.

13 Tesla, *My Inventions, 33.*

14 Tesla, *My Inventions, 33.*

15 Cheney, *Tesla, p. 131*

16 *Ibid.*, 40.

17 Carlson, *Tesla, 46.*

18 Cheney, *Tesla, 40.*

19 Carlson, *Tesla, 50.*

20 Tesla, *My Inventions, 35.*

21 Cawthorne, *Tesla, 141.*

22 Carlson, *Tesla, 247.*

23 *Ibid.*, 132.

24 *Ibid.*, 240.

25 Petkovich, D. (1927). "A Visit to Nikola Tesla," *Politika*, April 27, 4.

26 Cheney, *Tesla, 115.*

27 Seifer, *Wizard, 414.*

28 Hunt, I. and Draper, W. (1964). *Lightning in His Hands: The Life Story of Nikola Tesla*, Belgrade: Tesla Museum, A-398.

29 Seifer, *Wizard, 131-32.*

30 Carlson, *Tesla, 292.*

31 *Ibid.*, 242.

32 *Ibid.*, 243.

33 *Ibid.*, 361-62.

34 Seifer, *Wizard, 414-15.*

35 Carlson, *Tesla, 65.*

36 *Ibid.*, 66.

37 Nigel Cawthorne, *Tesla: The Life and Times of an Electric Messiah* (New York: Chartwell, 2014), 24.

38 See, for example, Cheney, *Tesla, 48*, and John J. O'Neill, *Prodigal Genius: The Life of Nikola Tesla*

(Kempton, IL: Adventures Unlimited Press, 2008), 60.

39 Mark J. Seifer, *Wizard: The Life and Times of Nikola Tesla: Biography of a Genius* (New York: Citadel, 1998), 31.

40 Carlson, *Tesla*, 49, 69, note 31.

41 Seifer, *Wizard, 30.*

42 See, variously, Cheney, *Tesla*, 49; Carlson, *Tesla*, 69–70; Cawthorne, *Tesla*, 24–25.

43 Valone, Thomas (2002). *Harnessing the Wheelwork of Nature: Tesla's Science of Energy*, Adventures Unlimited Press: Kempton (IL), 53.

44 Tesla, *My Inventions*, 71.

45 Carlson, *Tesla*, 70.

46 *Ibid.*, 73 and note 42.

47 Cawthorne, *Tesla*, 30.

48 Carlson, *Tesla*, 73.

49 Cheney, *Tesla*, 57.

50 Seifer, *Wizard*, 42–43.

51 *Ibid.*, 46.

52 *Ibid.*, 47.

53 Elihu Thomson, System of Electric Distribution. U.S. Patent 335,159, filed March 19, 1883, and issued February 2, 1886.

54 Elihu Thomson, "Novel Phenomenon of Alternating Currents," paper presented before the American Institute of Electrical Engineers, May 18, 1887.

55 Jack Foran, "The Day They Turned the Falls On: The Invention of the Universal Electric Power System," University of Buffalo Library Project Cases, 2013, http://library.buffalo.edu/libraries/projects/cases/niagara.htm.

56 Seifer, *Wizard*, 133–134.

57 "Nikola Tesla and His Wonderful Discoveries," *New York Herald*, April 23, 1893.

58 J. F. Patten, "Nikola Tesla," *Electrical World*, April 14, 1894, 489.

59 W. T. Stephenson, "Electric Light for the Future," *Outlook*, March 9, 1895.

60 Seifer, *Wizard, 115.*

61 *Ibid,,* 148.

62 "Is Tesla to Signal the Stars?" *Electrical World*, April 4, 1896.

63 Tesla, *My Inventions*, 48, 50.

64 *Ibid.*, 50.

65 Cheney, *Tesla*, 144.

66 Seifer, *Wizard, 183.*

67 *Ibid.*, 191.

68 "Tesla at 79 Discovers New Message Wave," *Brooklyn Eagle*, July 11, 1935.

69 Seifer, *Wizard*, 210.

70 *Ibid.*, 211.

71 *Ibid.*, 213, 218.

72 Carlson, *Tesla*, 264.

73 Seifer, *Wizard*, 214.

74 Cawthorne, *Tesla*, 82.

75 *Ibid.*, 86.

76 Seifer, *Wizard*, 220.

77 "Talking with the Planets," *Colliers*, February 9, 1901.

78 Julian Hawthorne, "And How Will Tesla Respond to Those Signals From Mars?" *Philadelphia North American*, 1901.

79 "Mr. Tesla's Science," *Popular Science Monthly*, February 1901, 436–37.

80 Seifer, *Wizard*, 223.

81 *Ibid.*, 244.

82 *Ibid.*, 238.

83 Peter Krass, "He Did It! (The Creation of U.S. Steel by J. P. Morgan)," *Across the Board 38*, no. 3 (May 2001): 27.

84 Seifer, *Wizard* (1989), 248.

85 *Ibid.*, 249.

86 *Ibid.*, 250.

87 *Ibid.*, 252. As far as I can deduce, Seifer constructs this conversation on the basis of several letters sent between Tesla and J. P. Morgan between 1900 and 1904.

88 *Ibid.*, 254-55.

89 Carlson, *Tesla*, 316.

90 *Ibid.*, 319.

91 Seifer, *Wizard*, 261.

92 Carlson, *Tesla*, 323.

93 *Ibid.*, 333–4.

94 *Ibid.*, 344.

95 *Ibid.*, 345.

96 Seifer, *Wizard*, 300.

97 Carlson, *Tesla*, 357.

98 *Ibid.*, 357–58.

99 *Ibid.*, 353.

100 *Ibid.*, 355.

101 *Ibid.*, 360.

102 Letter from Tesla to Morgan, February 17, 1905, as referenced in Carlson, *Tesla*, 361, note 70.

103 Seifer, *Wizard*, 323.

104 Carlson, *Tesla*, 371.

105 *Ibid.*, 372–73.

106 *Ibid.*, 374–75.

107 Seifer, *Wizard*, 401

108 Carlson, *Tesla*, 377.

109 *Ibid.*, 376.

110 *Ibid.*, 377.

111 Seifer, *Wizard*, 390.

112 *Ibid.*, 391.

113 Margaret Cheney and Robert Uth, *Tesla: Master of Lightning* (New York: Barnes & Noble, 1999), 127.

114 Seifer, *Wizard*, 370.

115 *Ibid.*, 397.

116 *Ibid.*, 398.

117 Cheney and Uth, *Lightning*, 125.

118 Carlson, *Tesla*, 379.

119 W. H. Secor, "Tesla's View on Electricity and the War," *Electrical Experimenter 5* (August 1917).

120 Carlson, *Tesla, 379.*

121 "Tesla at 78 Bares New 'Death Beam,'" *New York Times*, July 11, 1934.

122 Carlson, *Tesla*, 381–82.

123 Carlson, *Tesla*, 379.

124 Cheney and Uth, *Lightning*, 135.

125 Seifer, *Wizard*, 443.

126 Carlson, *Tesla*, 389, documents at least two such incidents, one in which Tesla instructs a young admirer to contact the dead J. P. Morgan and another in which he attempted to send a telegraph to Mark Twain, who had died 29 years earlier.

127 Cheney and Uth, *Lightning*, 133.

128 Seifer, *Wizard*, 443.

129 Carlson, *Tesla*, 389.

130 Fiorello La Guardia, "Eulogy to Nikola Tesla," presented on New York radio, January 10, 1943, http://www.teslasociety.com/eulogy.htm.

131 "Nikola Tesla Dead," *New York Sun*, January 1943.

CHAPTER 3

1 Arran Frood, "Riddle of Baghdad's Batteries," *BBC News*, February 27, 2003, http://news.bbc.co.uk/2/hi/science/nature/2804257.html.

2   P.T. Keyser, "The Purpose of the Parthian Galvanic Cells: A First-Century A.D. Electric Battery Used for Analgesia," *Journal of Near Eastern Studies* 52, no. 2 (April 1993), 81–98, http://personalpages.to.infn.it/~bagnasco/Keyser1993.pdf.

3   Keyser, "Galvanic Cells," 83.

4   Bruno Maddox, "Three Words That Could Overthrow Physics: 'What Is Magnetism?'" *Discover* (May 2008), http://discovermagazine.com/2008/may/02-three-words-that-could-overthrow-physics.

5   *Encyclopedia Britannica*, 11th ed. (vol. 2), s.v. "Arago, Dominique François Jean."

6   J. J. O'Connor and E. F. Robertson, "Michael Faraday," School of Mathematics and Statistics, University of St. Andrews, Scotland, May 2001, http://www-history.mcs.st-andrews.ac.uk/Biographies/Faraday.html.

7   Nicholas Gerbis, "How Induction Cooktops Work," How Stuff Works, last modified December 9, 2009, accessed September 17, 2014, http://home.howstuffworks.com/induction-cooktops2.htm.

8   I am indebted to Dr. Alexandra "Sascha" von Meir, co-director of the California Institute for Energy and Environment's Electric Grid project, for many of the analogies used to express electrical phenomenon. For a more comprehensive explanation of electricity basics, see von Meier, *Electrical Power Systems: A Conceptual Introduction* (Hoboken, NJ: John Wiley & Sons, 2006).

CHAPTER 4

1   "The Importance of Electric Motor Drives," What-When-How.com, The-Crankshaft Publishing, http://what-when-how.com/electric-motors/the-importance-of-electric-motor-drives.

2   Harry Bruinius, "A Superstorm Sandy Legacy: Gas Pumps That Work When Power Is Out," *The Christian Science Monitor*, October 28, 2013, http://www.csmonitor.com/Environment/2013/1028/A-superstorm-Sandy-legacy-Gas-pumps-that-work-when-power-is-out.

3   Zachary Shahan, "Electric Motors Use 45% of Electricity, Europe Responding," CleanTechnica.com, June 16, 2011, http://cleantechnica.com/2011/06/16/electric-motors-consume-45-of-global-electricity-europe-responding-electric-motor-efficiency-infographic.

4   "Tesla's AC Induction Motor Is One of the Ten Greatest Discoveries of All Time," Tesla Memorial Society of New York, Hall of Fame, http://www.teslasociety.com/hall_of_fame.htm.

5   Aleksandar Culibrk, "Most Significant Inventions," Tesla 150th Anniversary: Liberating Energy, http://www.b92.net/eng/special/tesla/life.php?nav_id=36495.

6   Nikola Tesla, Electro-Magnetic Motor, U.S. Patent 381,968, filed October 12, 1887, and issued May 1, 1888.

7   Alexandra von Meier, *Electric Power Systems: A Conceptual Introduction* (Hoboken, NJ: John Wiley & Sons, 2006), 85.

8   R. Victor Jones, "Samuel Thomas von Sömmering's 'Space Multiplexed' Electrochemical Telegraph (1808–1810)," *From Semaphore to Satellite* (Geneva: International Telecommunications Union), http://people.seas.harvard.edu/~jones/cscie129/images/history/von_Soem.html.

9   See David Nye, *Electrifying America: Social Meanings of New Technology* (Cambridge, MA: MIT Press, 1990).

10   O'Connor and Robertson, "Michael Faraday."

11   "Hippolyte Pixii Biography (1808–1835)," How Products are Made: Geradrus Mercator to James Eumsey, http://www.madehow.com/inventorbios/91/Hippolyte-Pixii.html.

12   Carlson, *Tesla*, 37.

13   "Magneto-electric machine by Pixii," Museo Galileo, http://catalogue.museogalileo.it/object/MagnetoelectricMachineByPixii.html.

14 Tesla, *My Inventions*, 33.

15 Carlson, *Tesla*, 44–5.

16 Tesla, *My Inventions*, 33.

17 *Ibid., 33.*

18 Cheney, *Tesla*, 43.

19 Carlson, *Tesla*, 50.

20 Tesla, *My Inventions*, 35.

21 *Ibid., 35.*

22 Carlson, *Tesla*, 52.

23 *Ibid., 52* footnote 42.

24 Seifer, *Wizard*, 28.

25 Carlson, *Tesla*, 51 footnote 40.

26 Tesla, *My Inventions*, 37.

27 Seifer, *Wizard*, 63.

28 *Ibid., 101.*

29 Walter Baily, "A Mode of Producing Arago's Rotation," *Philosophical Magazine: A Journal of Theoretical, Experimental and Applied Physics* (June 28, 1879) 286.

30 Judge William Kneeland Townsend, Westinghouse Electric & Manufacturing Company v. New England Granite Company, 103 F. 951, 36 (August 29, 1900).

31 Marcel Deprez "On the Electrical Synchronism of Two Relative Motions and Its Application to the Construction of a New Electrical Compass," *Reports of the French Academy of Sciences* (1881).

32 Townsend, *New England Granite*, 31.

33 *Id.*, 40.

34 *Id.*, 44 (emphasis added).

35 Carlson (2013), 96.

36 Nikola Tesla, Electrical Transmission of Power, U.S. Patent 511,559; and System of Electrical Power Transmission, U.S. Patent 511,560; both filed December 8, 1888, and issued December 26, 1893.

37 Carlson, *Tesla*, (2013), 97.

38 *Ibid., 97–8.*

39 *Ibid., 98.*

40 *Ibid., 98.*

41 Judge Albert Thompson, Westinghouse Electric & Manufacturing Company v. Dayton Fan & Motor Company, 106 F. 724, 6–7 (1901).

42 *Id.*, 13.

43 Judge Henry Franklin Severens, Dayton Fan & Motor Company v. Westinghouse Electric & Mfg. Company, 118 F. 562 (1902).

44 *Id.*, 25.

45 *Id.*, 33.

46 Judge William Kneeland Townsend, Westinghouse Electric & Manufacturing Company v. Catskill Illuminating & Power Company, 121 F. 831 (February 25, 1903).

47 *Complete Dictionary of Scientific Biography*, s.v. "Galileo Ferraris," http://www.encyclopedia.com/topic/Galileo_Ferraris.aspx.

48 "Galileo Ferraris: Physicist, Pioneer of Alternating Current Systems," Engineering Hall of Fame, Edison Tech Center, 2011, http://www.edisontechcenter.org/GalileoFerraris.html.

49 "Biography of Galileo Ferraris," Great Scientists, Incredible People: Biographies of Famous People, 2013, http://www.incredible-people.com/biographies/galileo-ferraris.

50 Carlson, *Tesla*, 104.

51 *Ibid.*, 107.

52 *Ibid.*, 104.

53 Townsend, *Catskill*, 9.

54 *Ibid., 11.*

55 Judge Colt, Westinghouse Electric & Manufacturing Company v. Stanley Instrument Company, District Court, Massachusetts (March 11, 1903).

56 Judge John Raymond Hazel, Westinghouse Electric & Manufacturing Company v. Mutual Life Insurance Company, 129 F. 213 (1904).

57 *Id.*, 215.

58 *Id.*, 216.

59 *Id.*, 217.

60 *Id.*, 218.

61 *Id.*, 219.

62 *Id.*, 219.

63 *Id.*, 219.

64 Seifer, *Wizard*, 51.

65 Carlson, *Tesla*, 112–13.

66 Seifer, *Wizard*, 49–50.

67 *Ibid., 50.*

68 Hazel, *Mutual Life*, 216.

69 Carlson, *Tesla*, 131.

70 Seifer, *Wizard*, 66 note 2.

71 William Stanley, "The Expiration of the Tesla Patents, " *Electrical World and Engineer*, May 6, 1905.

72 Carlson, *Tesla*, 130.

73 *Ibid., 130.*

74 *Ibid., 129–30.*

75 Seifer, *Wizard*, 150.

76 *Ibid., 178.*

77 Ronald Schatz, *The Electrical Workers: A History of Labor at General Electric and Westinghouse, 1923–60* (Urbana: University of Illinois Press, 1983), 5.

78 William Stanley. "Expiration of the Tesla Split-Phase Patents," *Electric World and Engineer*, July 7–December 29, 1910.

79 Stanley, "Tesla Patents," 828.

80 Seifer, *Wizard*, 163.

81 *Ibid.*, 266.

82 Carlson, *Tesla*, 239.

83 Stanley, "Tesla Patents," 828.

CHAPTER 5

1 *Encyclopedia Britannica*, s.v. "Joseph Henry," accessed October 5, 2014, http://www.britanicca.com/EBchecked/topic/261387/Joseph-Henry.

2 Nicholas Callan, "A Description of an Electromagnetic Repeater, or of a Machine by Which the Connection between the Voltaic Battery and the Helix of an Electromagnet May Be Broken and Renewed Several Thousand Times in the Space of One Minute," *Sturgeon's Annals of Electricity* 1 (1837), 229–30, http://books.google.come/books?id=SXgMAAAAYAAJ&pg=PA@@(&lpg=PA229#v=onepage&q&f=false.

3 A. Frederick Collins, *The Design and Construction of Induction Coils* (New York: Munn & Co., 1908), 98.

4 Amin Sayed Saad, "Transformer Theory," *Electrical Power System and Transmission Network*, accessed October 15, 2014, http://www.sayedsaad.com/fundmental/index_transformer.htm,.

5 J. C. Maxwell, "A Dynamical Theory of the Electromagnetic Field," *Philosophical Transactions of the Royal Society of London* 155 (1865), 459–512, http://upload.wikimedia.org/wikipedia/commons/1/19/A_Dynamical_Theory_of_the_Electromagnetic_Field.pdf.

6 J. J. O'Connor, and E. F. Robertson, "James Clerk Maxwell," School of Mathematical and Computational Sciences, University of St. Andrews, November 1997, http://web.archive.org/web/20110128034939/http://www-groups.dcs.st-and.ac.uk/~history/Biographies/Maxwell.html.

7 Heinrich Hertz, "On Electromagnetic Waves in Air and Their Reflection," *Weidemann's Annual* 34 (1888): 610.

8 Cheney and Uth, *Lightning*, 35–6.

9 Nikola Tesla, Electrical Transformer or Induction Device, U.S. Patent 433,702, filed March 26, 1890, and issued August 5, 1890; System of Electric Lighting, U.S. Patent 454,622 A, filed

April 25, 1891, and issued June 23, 1891; Electro Magnetic Motor, U.S. Patent 455,067 A, filed January 27, 1891, and issued June 30, 1891; Method of and Apparatus for Electrical Conversion and Distribution, U.S. Patent 462,418 A, filed February 4, 1891, and issued November 3, 1891; Electro Magnetic Motor, U.S. Patent 464,666 A, filed July 13, 1891, and issued December 8, 1891 .

10 Edvard Csanyi, "What Is the Best Transformer Coolant?" *Electrical Engineering Portal, April 2012,* http://electrical-engineering-portal.com/what-is-the-best-transformer-coolant.

11 von Meier, *Electric,* 172.

12 "Sulfur Hexafluoride," *Science Daily,* 2014, http://www.sciencedaily.com/articles/s/sulfur_hexa-fluoride.htm.

13 Slingo, W. and Brooker, A. (1900). *Electrical Engineering for Electric Light Artisans,* London: Longmans, Green & Company, 607. http://www.worldcat.org/title/electrical-engineering-for-electric-light-artisans-and-students-embrac-ing-those-branches-prescribed-in-the-syllabus-is-sued-by-the-city-and-guilds-technical-institute/oclc/264936769.

14 Gribben, J. (2004). *The Scientists: A history of science told through the lives of its greatest inventors,* Random House, 424-32.

15 ___ (1896). "Mr. Moore's Etheric Light: The Young Newark Electrician's New and Successful Device," *New York Times,* October 2. http://query.nytimes.com/gst/abstract.html?res=9400E1DE-133BEE33A25751C0A9669D94679ED7CF.

16 ___ (2013). "Who invented the fluorescent light bulb?" *Answers.com.* http://wiki.answers.com/Q/Who_invented_the_fluorescent_light_bulb.

17 M.W. (2010). "Who invented the fluorescent lamp? Myths about Nikola Tesla and Agapito Flores," *Edison Tech Center.* http://www.edisontech-center.org/WhoInventedFluorLamp.html.

18 Encyclopedia Brittanica (2013). "Peter Cooper Hewitt," *Encyclopaedia Britannica.* http://www.britannica.com/EBchecked/topic/264522/Peter-Cooper-Hewitt?anchor=ref196631

19 Carlson, *Tesla,* 123–24.

20 Cawthorne, *Electric Messiah,* 55.

21 Carlson, *Tesla,* 124.

22 Cheney and Uth, *Lightning,* 45.

23 Tesla, *My Inventions,* 56.

24 Cheney and Uth, *Lightning,* 45.

25 Tesla, *My Inventions,* 50.

26 Nikola Tesla, Electrical Transformer or Induction Device, U.S. Patent 433,702, filed March 26, 1890, and issued August 5, 1890.

27 ———, Method and Apparatus for Electrical Conversion and Distribution, U.S. Patent 462,418 A, filed February 4, 1891, and issued November 3, 1891.

28 *Ibid.*

29 Nikola Tesla, Coil for Electro Magnets, U.S. Patent 512,340 A, filed July 7, 1893, and issued January 9, 1894. http://www.google.com/patents/US512340.

30 Carlson, *Tesla,* 144.

31 *Ibid., 87.*

32 *Ibid., 88.*

33 *Ibid., 88.*

34 *Ibid., 61.*

35 *Ibid,* 61. Although the exact dates may not be accurate, Carlson claims that Tesla came upon a broken ring transformer (probably used to power a Ganz arc lighting system) in a corner of Zipernowsky's workshop sometime in early 1882, before Ottó Bláthy began working there.

36 "Otto Titusz Blathy," *Biographies of Famous People,* http://incredible-people.com/biographies/otto-titusz-blathy/.

37 Carlson, *Tesla*, 60. *[See* Osana Mario, "Historische Betrachtungen über Teslas Erfindungen des Mehrphasenmotors und der Radiotechnik um die Jahrhundertwende," in *Nikola Tesla-Kongress für Wechsel-und Drehstromtechnik,* proceedings of a conference held at the Technical Museum in Vienna, 6–13 September 1953, (Vienna: Springer-Verlag), 6–9.]

38 Seifer, *Wizard,* 27.

39 Carlson, *Tesla,* 145.

40 Seifer, *Wizard,* 83.

41 Carlson, *Tesla,* 146.

42 Seifer, *Wizard,* 86.

43 "Fleming, John Ambrose (FLMN877JA)," *A Cambridge Alumni Database, University of Cambridge,* http://venn.lib.cam.ac.uk/cgi-bin/search.pl?sur=&suro=c&fir=&firo=c&cit=&cito=c&c=all&tex=%22FLMN887JA%22&sye=&eye=&col=all&maxcount=50.

44 Seifer, *Wizard,* 90 note 33.

45 William Spottiswoode, "Description of a Large Induction Coil," *The London, Edinburgh, and Dublin Philosophical Magazine* 3 no. 15 (January 1877), 30.

46 "Leyden Jar," How Stuff Works, http://science.howstuffworks.com/leyden-jar-info.htm.

47 Stone, J., *Marconi Wireless Telegraph Company of America v. United States,* 320 U.S. 1 (1943), 53-54

48 Seifer, *Wizard,* 92.

49 *Ibid.,* 95.

50 *Ibid.,* 95.

51 Nitum, "Biography of Otto Titusz Blathy," nitum.wordpress.com, October 5, 2012, https://nitum.wordpress.com/tag/Bláthys-other-inventions-include-the-induction-meter/.

52 Seifer, *Wizard,* 95.

53 Weinstein, E. (2007). "Hemholtz, Hermann von (1821-1894)," *ScienceWorld.* http://scienceworld.wolfram.com/biography/Helmholtz.html.

54 See Heinrich Hertz and Daniel Evan Jones, *Electric Waves: Being Researches in the Propagation of Electric Action with Finite Velocity Through Space* (London: Macmillan & Co, 1893).

55 Seifer, *Wizard,* 95–6.

Chapter 6

1   Brown, William C. (1984). "The History of Power Transmission by Radio Waves," *IEEE Transactions on Microwave Theory and Technique,* v.32:9, September, 1236.

2   Brown, *Transactions,* 1234.

3   *Id.,* 1231.

4   *Ibid.*

5   Jaeger, M. (2012). "Tesla and wireless energy: the power that could have been," *Washington Times,* July 15. http://communities.washingtontimes.com/neighborhood/energy-harnassed/2012/jul/15/GreaterThanEnergy_TeslaWireless_1000/.

6   *Ibid.*

7   Schroeder, H. (1923). *History of Electric Light,* Washington, D.C.: Smithsonian Institution, foreword. http://archive.org/stream/historyofelectri00schr/historyofelectri00schr_djvu.txt.

8   Blanchard, J. (1941). "History of Electrical Resonance," *Bell System Technical Journal,* v.20:4, October, 419. https://archive.org/details/bstj20-4-415. According to Blanchard, Michael Pupin recounted the story during a meeting of the American Institute of Electrical Engineers.

9   *Ibid.,* 420-1.

10 Sibakoti, M. J. and Hambleton, J. (2011). *Wireless Power Transmission Using Magnetic Resonance,* Cornell: Ithaca, NY, December, 1. http://www.academia.edu/8143067/Wireless_Power_Transmission_Using_Magnetic_Resonance.

11 Carlson, *Tesla,* 178-9.

12 *Ibid.,* 179.

13 *Id.*, 209.

14 *Ibid.*

15 Prescott, G. B. (1860). *History, Theory and Practice of the Electric Telegraph*, 398-400. http://earlyradiohistory.us/1860stei.htm.

16 *Ibid.*

17 Carlson, *Tesla*, 211.

18 *Id.*, 251.

19 *Id.*, 189.

20 *Ibid.*

21 *Id.*, 194-95.

22 *Id.*, 207.

23 Seifer, *Wizard*, 140.

24 Carlson, *Tesla*, 218-19.

25 *Ibid.*, 225-27.

26 U.S. Patent 645,576 A. "System of transmission of electrical energy," filed September 2, 1897; granted March 20, 1900. http://www.google.com/patents/US645576.

27 *Ibid.*

28 *Id.*, 2.

29 *Id.*

30 *Id.*

31 *Id.*

32 Carlson, *Tesla*, 259.

33 ___ (1899). "Tesla Says…" *New York Journal*, April 30.

34 Carlson, *Tesla*, 188-89.

35 *Ibid.*

36 Cheney, *Tesla*, 177.

37 *Ibid.*, 187-88.

38 *Ibid.*

39 Carlson, *Tesla*, 300-01.

40 Tesla, *My Inventions*, 56.

41 *Ibid.*, 52.

42 U.S. Patent 649,621. "Apparatus for transmission of electrical energy," filed February 19, 1900, specification forming part of Letters Patent No. 649,621 dated May 15, 1900. http://www.google.com/patents/US649621.

43 *Ibid.*, 1.

44 See Cheney and Uth, *Master of Lightning*, 67.

45 Bethune, B. (2008). "Did Bell Steal the Idea for the Phone (Book Review)," *Maclean's, February 4.* http://www.the-canadianencyclopedia.com/en/article/did-bell-steal-the-idea-for-the-phone-book-review.

46 *Ibid.*

47 Thompson, S. P. (1883). Phillip Reis: inventor of the telephone: A biographical sketch, with documentary testimony, translations of the original papers of the inventor and contemporary publications, *London: E. & F.N.* Spon., 182. http://www.archive.org/stream/philippreisisinven00thomrich_djvu.txt.

48 Munro, J. (1883). *Heroes of the Telegraph*, 216. http://books.google.com/books?id=XtMmww1xGqkC.

49 ___(2003). "Bell 'did not invent the telephone'", BBC News – Science/Nature, December 1. http://news.bbc.co.uk/1/hi/sci/tech/3253174.stm)

50 Thompson, *Phillip Reis.*

51 Legat, V. (1862). *Reproducing sounds on extra galvanic way, Rutgers, The Thomas Edison Papers,* Document #TI2459, Litigation Series. Http://edison.rutgers.edu/NamesSearch/glocpage.php3?gloc=TI2&.

52 Casson, H. N. (1910). *The History of the Telephone*, Chicago: McClurg., 95. http://inventors.about.com/gi/dynamic/offsite.htm?site=http://etext.lib.virginia.edu/toc/modeng/public/CasTele.html.

53 *Ibid.*, 96.

54 ___ (2014). "Timeline of the Telephone," Wikipedia, October 29. http://en.wikipedia.org/wiki/Timeline_of_the_telephone.

55 Shulman, S. (2009). *The Telephone Gambit: Chasing Alexander Graham Bell's Secret*. New York: W.W. Norton & Company, 125.

56 *Ibid.*

57 U.S. Patent 350,299. "Mode of Electrical Communication," filed March 24, 1882, granted October 5, 1886. http://www.google.com/patents/US350299.

58 Berman, B. (2014). *Zoom: From Atoms and Galaxies to Blizzards and Bees: How Everything Moves*. Google eBook., June 24.

59 *American Bell Telephone Company v. Amos E. Dolbear et. al.*, Circuit Court of the United States, District of Massachusetts, Bill of Complaint, Doc. #1626, filed October 10, 1881, 10. https://archive.org/stream/americanbelltel00masgoog#page/n22/mode/2up.

60 Gray, J. (1883). *American Bell Telephone Company v. Amos E. Dolbear et. al.*, Circuit Court of the United States, District of Massachusetts, Opinion of the Court, January 24, 5-6. https://archive.org/stream/americanbelltel00masgoog#page/n506/mode/2up.

61 *Ibid.*, 6.

62 ___(1881). "A New Telephone System," *Scientific American*, June 18. http://www.mchine-history.com/DolbearTelephonicSystem1881.

63 *Ibid.*

64 Basilio, C. (2003). "Antonio Meucci inventore del telefono," *Notiziario Tecnico Telecom Italia*, v.12: 1, December, 109. http://www.museoaica.it/Museo_Aica/esplora/fili/pdf/12_Muecciday3.pdf.

65 *Ibid.*

66 Meucci, S. (2010). *Antonio Meucci and the Electric Scream: The Man who Invented the Telephone*, Boston: Branden Books, 72-73.

67 *Ibid.*

68 Meucci, A. (1871). "Sound Telegraph," U.S. Patent Caveat No. 3335, filed December 28, 1871. http://files.meetup.com/202213/MeucciMarch07.pdf.

69 *American Bell Telephone Co. v. Globe Telephone Co.*, 31 Fed. Rep. 729 (Circuit Court, S. D. New York, July 19, 1887).

70 Kennedy, R. C. (2001). "On this Day," *New York Times on the Web Learning Network*. http://www.nytimes.com/learning/general/onthisday/harp/0213.html.

71 ___(2013). "Elisha Gray," *IEEE Global History Network*. http://www.ieeeghn.org/wiki/index.php/Elisha_Gray.

72 Bellis. M. (2014). "Elisha Gray's Patent Caveat: Transmitting Vocal Sounds Telegraphically," *About.com – Famous Inventors*. http://inventors.about.com/od/gstartinventors/a/Elisha_Gray_2.htm.

73 Shulman, S. (2008). *The Telephone Gambit*. New York: W.W. Norton, 71.

74 Baker, B. H. (2000). *The Gray Matter: The Forgotten Story of the Telephone*. St. Joseph, MI: Telepress, A43-A44.

75 Evenson, A. E. (2000). *The Telephone Patent Conspiracy of 1876: The Elisha Gray – Alexander Bell Controversy*, North Caroline: McFarland, 167-171.

76 Southwest Museum of Engineering, Communications and Computation (SMECC) (2007). "Mahlon Loomis – First Wireless Telegrapher," *Mahlon Loomis*. http://www.smecc.org/mhlon_loomis.htm.

77 *Ibid.*

78 Compare SMECC to Bagg, E. N. (1913). "Wireless Telegraph's Pioneer," *Western New England Advertiser*, v.3, 27.

79 SMECC, *Mahlon Loomis*.

80 Loomis, M., U.S. Patent 129,971 A. "Improvement in telegraphing," filed July 30, 1872, 1. http://www.google.com/patents/US129971.

81 *Ibid.*

82 Ward, L.W., U.S. Patent 126,356 A. "Improvement in collecting electricity for

telegraphing," filed April 30, 1872. http://www.google.com/patents/US126356.

83 *Ibid.*, 2.

84 Loomis, *Improvement in Telegraphing*, 1.

85 Sauer, A., et. al. (2000). "Dolbear, Amos Emerson, 1837-1910," *Concise Encyclopedia of Tufts History*, Tufts Digital Library, http://dl.tufts.edu/catalog/tei/tufts:UA069.005.DO.00001/chapter/D00047.

86 Dolbear, A., *Tesla*, 139.

87 U.S. Patent 350,299. "Mode of electric communication," filed March 24, 1882, granted October 5, 1886. http://www.google.com/patents/US350299.

88 *Ibid.*, 2.

89 Carslon, *Tesla*, 139.

90 Cheney, *Tesla*, 259.

91 Secor, H. W. (1917). "Tesla's Views on Electricity and the War," *The Electrical Experimenter*, v.5:4, August, 229. http://electrical-experimenter.com/n4electricalexperi05gern.pdf.

92 Cheney, *Tesla, 265.*

93 Kostenko, A. A., et. al. (2001). "Radar prehistory, Soviet side: three-coordinate L-band pulse radar developed in Ukraine in the late 30's," before the *Antennas and Propagation Society International Symposium*, IEEE, v.4, July 8-13, 44-47.

94 Holmann, M. (2007). "Christian Hulsmeyer, the inventor," Radar World. http://www.radar-world.org/huelsmeyer.html.

95 German Patent DE165546 (April, 1904); British Patent 16556 (September 23, 1904).

96 See Watson, R. C. (2009). *Radar Origins Worldwide: History of Its Evolution in 13 Nations Through World War II*, Tafford Publishing.

97 Japanese Patent #69115. http://www.aktuellum.com/circuits/antenna-patent/patents/69115-A.gif.

98 Yagi, H. and Uda, S. (1926). "Projector of the Sharpest Beam of Electrical Waves," *Proceedings of the Imperial Academy*, v.2:2, 49-52. https://www.jstage.jst.go.jp/article/pjab1912/2/2/2_2_49/_article.

99 IEEE Antennas and Propagation Society (1998). *International Symposium of the Institute of Electrical and Electronics Engineers, National Radio Science Meeting*, Atlanta. 26-27. http://books.google.com.my/books?cd=8&id=MgpWAAAAMAAJ&dq=yagi+singapore&focus=searchwithinvolume&q=yagi+singapore.

CHAPTER 7

1 Brittany Hayes,"Innovation & Infringement: The Wright Brothers, Glenn H. Curtiss, and the Aviation Patent Wars," U.S. History Scene, June 7, 2012, http://www.ushistoryscene.com/uncategorized/wrightbrotherspatentwars; Eleanor Knowles, "First Manned Flight, Brompton Dale," Engineering Timelines, accessed January 2015, http://www.engineering-timelines.com/scripts/engineeringItem.asp?id=546.

2 Victoria Wollaston, "The Wright Brothers Were NOT the First to Fly a Plane . . . ," *Daily Mail*, May 20, 2013, http://www.dailymail.co.uk/sciencetech/article-2327286/The-Wright-Brothers-NOT-fly-plane--German-pilot-beat-years-earlier-flying-car-claims-leading-aviation-journal.html.

3 Joe Nocera, "Greed and the Wright Brothers," *New York Times*, April 18, 2014, http://www.nytimes.com/2014/04/19/opinion/nocera-greed-and-the-wright-brothers.html?_r=0.

4 Darin Gibby, *Why Has America Stopped Inventing?* (New York: Morgan James Publishing, 2011), 192.

5 *Ibid.*, 193.

6 Carlson, *Tesla*, 397–98.

7 Inez Hunt and Wanetta Draper, *Lightning in His Hand: The Life Story of Nikola Tesla* (Thousand Oaks, CA: SAGE, 1964).

8 Carlson, *Tesla*, 398.

9 *Ibid.*, 409–10.

10 *Ibid.*, 411.

11 *Ibid.*, 413.

12 *Ibid.*, 402.

13 *Ibid.*, 87.

14 *Ibid.*, 402.

15 Philip Torchio, "Distributing Systems from the Standpoint of Theory and Practice," *Transactions of the International Electrical Congress* (1905): 573.

16 David Rushmore and Eric Lof, *Hydro Electric Power Stations*, 2nd ed. (New York: John Wiley & Sons, 1923), 5.

17 Carlson, *Tesla*, 402.

18 J. Wiley Blount Rutledge (dissenting in part), *Marconi Wireless Telegraph Co. of America v. United States* (1943), 320 U.S. 1, 75.

19 *Id.*, 65–66.

20 Thomas Kuczmarski, "Innovation Always Trumps Invention," *Businessweek*, January 19, 2011, http://www.businessweek.com/innovate/content/jan2011/id20110114_286049.htm.

21 Tim Worstall, "Using Apple's iPhone to Explain the Difference Between Invention and Innovation," *Forbes*, April 20, 2014, http://www.forbes.com/sites/timworstall/2014/04/20/using-apples-iphone-to-explain-the-difference-between-invention-and-innovation/.

22 Lemley, *Michigan Law Review*, 714.

23 *Ibid.*

24 Carlson, *Tesla*, 122.

25 See Marc Seifer, *Transcending the Speed of Light: Consciousness, Quantum Physics & the Fifth Dimension* (New York: Inner Traditions, 2008).

26 David Jerale, "Myths and Rumors Persist in the Tale of Legendary Inventor Nikola Tesla," *The Libertarian Republic*, September 12, 2013, http://thelibertarianrepublic.com/evil-capitalists-prevent-nikola-tesla-creating-free-energy/#.VMKMZyvF8sw.

27 Carlson, *Tesla*, 363.

28 *Ibid.*, 406–408.

29 Carlson, *Tesla*, 406.

INDEX

Page references in *italics* indicate photographs or illustrations.

(No Model.)

N. TESLA.
DYNAMO ELECTRIC MACHINE.

No. 390,721.   Patented Oct. 9, 1888.

Exciter

Generator

Motor

Transformer

WITNESSES:

Raphaël Netter

Robt. F. Gaylord

INVENTOR

Nikola Tesla

BY

Duncan, Curtis &

Page ATTORNEYS.

PHOTO CREDITS

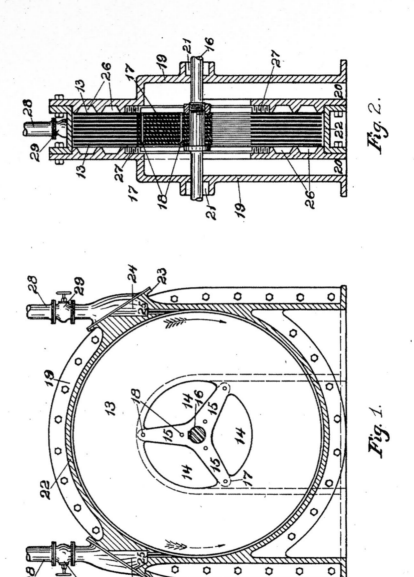

Fig. 2.

Fig. 1.

Witnesses:

R. Diaz Buitrago

Wm Bohleber

Nikola Tesla, Inventor

By his Attorneys

Kerr Page Cooper & Hayward

ACKNOWLEDGMENTS

A major thesis of this book is that great inventions are rarely the product of single minds working in isolation. The same is true of most great books (at the very least, most nonfiction works). And while I leave it to the critics to proclaim it great or grotesque, *The Truth about Tesla* is no different. Every word, every image, and almost all thought contained herein are due in whole or in part to the tireless contributions of a multitude.

First and foremost, I must salute Jeff McLaughlin at Race Point Publishing (Quarto US) for sharing my vision of producing more than a simple, illustrated biography of Nikola Tesla and then demonstrating the patience and flexibility to allow me to bring that vision to life. There are few publishers left that can see beyond the bottom line far enough to support an endeavor for its intellectual worth, not just its commercial appeal, and Jeff is among the dwindling few.

Much praise should be bestowed upon Greg Oviatt and Erin Canning, who managed the many moving parts—from editing to indexing—that are required to make a book this complex appear so simple. For the photographs, many rare and available nowhere else in print, I must thank Stacey Stambaugh, who worked tirelessly to track down those of the highest quality. Heather Rodino, who copyedited the book, was an invaluable source of information and insight, and thanks to Meredith Hale, whose proofing was meticulous.

Any professor of basic electrical engineering understands the challenge of simplifying the science of electricity. Many of the most useful metaphors used in this book derive from Alexandra "Sacha" von Meier, co-director of the Electric Grid Program at the California Institute for Energy and the Environment. I am grateful to her for teaching me the engineering fundamentals of electricity transmission, and for serving as technical editor of this book.

Rob Pilaud and Filip Vanevski, for whom this book is dedicated, are two of the most talented patent agents in the field and provided quick and accurate answers to my many inquiries regarding patent law. Thanks are due as well to Stefanie Cangiano for alerting me to the intriguing account of Antonio Meucci, the inclusion of which added immensely to the section on the origins of the telephone.

Finally, I owe the greatest gratitude to the Tesla biographers whose works formed the basis of this book. Individually, the primary research they undertook over many years helped quickly direct me to the most useful sources and sped the development of this book. Collectively (and perhaps unwittingly), their inconsistencies signaled the best areas to explore for sources of some of the most persistent myths (as well as the kernels of truth necessary to dispel them). Chief among this venerable tribe is Marc J. Seifer, who also contributed the introduction. His seminal work, *Wizard: The Life and Times of Nikola Tesla*, a book whose development spanned nearly two decades of painstaking research, remains far-and-away the most comprehensive and accurate account of Tesla's life.

Margaret Cheney's renowned biography, *Tesla: Man Out of Time* (as well as her more recent biography with Robert Uth), proved an invaluable source for the most colorful stories from Tesla's life. *Tesla: Inventor of the Electrical Age*, the more recent biography by W. Bernard Carlson, provided not only the best technical explanations of Tesla's inventions that I could find, but also the most fodder for criticism. Our perspectives align far more than they conflict, and the greatest praise I could receive is his collegial reaction to this critique. His lifelong study of Tesla, and of the history of science and technology in general, is commendable and a career I can only hope to emulate.

No. 649,621.

Patented May 15, 1900.

N. TESLA.

APPARATUS FOR TRANSMISSION OF ELECTRICAL ENERGY.

(Application filed Feb. 19, 1900.)

(No Model.)

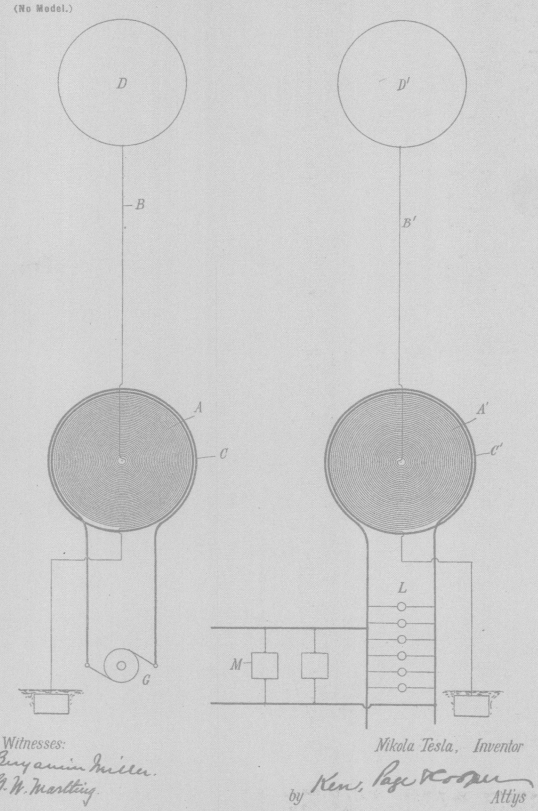

Witnesses:

Benjamin Miller.

G. W. Martling.

Nikola Tesla, Inventor

by Ken, Page & Cooper Attys